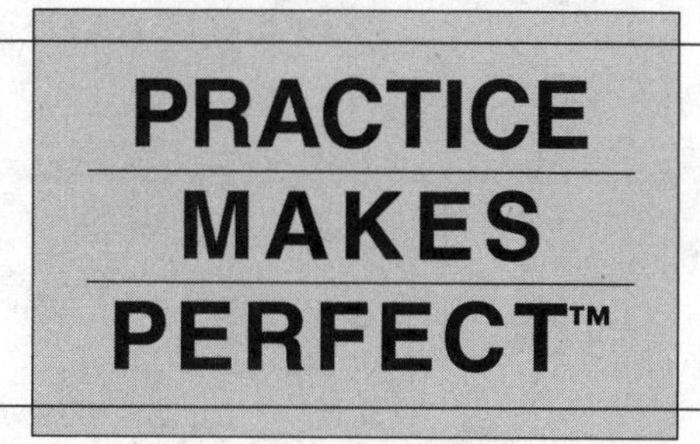

Algebra II

Algebra II

Review and Workbook

THIRD EDITION

Chris Monahan
Laura Favata

New York Chicago San Francisco Athens London Madrid Mexico City
Milan New Delhi Singapore Sydney Toronto

1 2 3 4 5 6 7 8 9 LHS 27 26 25 24 23 22

ISBN 978-1-264-28642-3
MHID 1-264-28642-2

e-ISBN 978-1-264-28643-0
e-MHID 1-264-28643-0

Interior design by Village Bookworks

Contents

Introduction

Algebra II uses the skills and concepts learned in Algebra I as well as many of the concepts learned in Geometry. You will learn many new concepts in Algebra II, the most important of which is the idea of a function. Functions are a fundamental building block for the development of higher mathematics. Questions about domain and range will appear in almost every chapter of the course. Using transformations to extend a basic function into a family of functions will help you develop a better understanding of functions and allow you to develop a mental image of the function before you look at the graph on your graphing calculator or computer. Mathematics is a tool used in nearly every aspect of the world of work. Examples for many of these applications are included in *Practice Makes Perfect Algebra II Review and Workbook*.

Whereas you can read a piece of literature or a document for a social studies class and then quietly contemplate the meaning of what you read, mathematics requires a more active approach. You should read the text and the examples provided for you. You should also do the guided exercises after you have finished reading a section to ensure you understand the steps involved in solving the problems. Once you have done this, then you should do the exercises at the end of the section. Check your answers with the answer key in the back of the book to verify you have done them correctly. Some problems require very little writing, and some require the use of technology such as a graphing calculator or computer software. Most of the exercises can be done with paper and pencil. Take the time to do all of the exercises. You will learn a great deal from the time and effort you put forth.

Algebra II Review and Workbook is written so that you can practice a few concepts at a time. This does not mean that the examples and exercises will not apply what you learned in a previous lesson or from other courses, but that the crux of the problem is to help you better understand the concept from that particular section. If you find that a prior concept is slowing you down, take the time to go back to that section to get a better understanding of what you are missing.

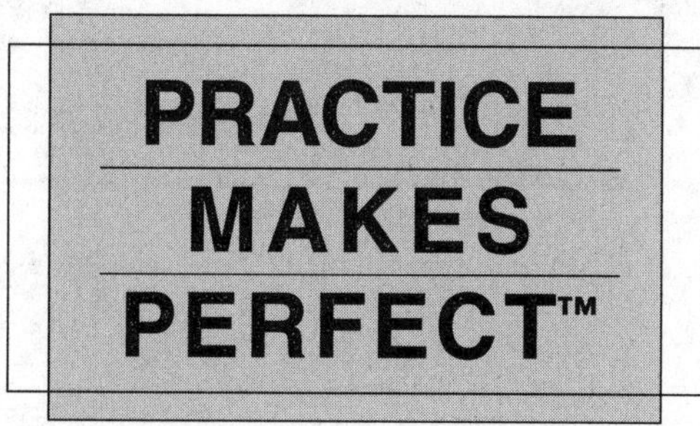

Algebra II

Functions: An introduction

Mathematics is known for its ability to convey a great deal of information with the use of a minimum number of symbols. While this may be initially confusing (if not frustrating) for the learner, the notation of mathematics is a universal language. In this chapter, you will learn about function notation.

Relations and inverses

One of the major concepts used in mathematics is relations. A **relation** is any set of ordered pairs. The set of all first elements (the input values) is called the **domain**, while the set of second elements (the output values) is called the **range**. Relations are traditionally named with a capital letter. For example, given the relation

$$A = \{(2, 3), (-1, 5), (4, -3), (2, 0), (-9, 1)\}$$

the domain of A (written D_A) is $\{-9, -1, 2, 4\}$. The domain was written in increasing order for the convenience of reading, but this is not required. The element 2, which appears as the input for two different ordered pairs, needs to be written only one time in the domain. The range of A (written R_A) is $\{-3, 0, 1, 3, 5\}$.

The **inverse** of a relation is found by interchanging the input and output values. For example, the inverse of A (written A^{-1}) is

$$A^{-1} = \{(3, 2), (5, -1), (-3, 4), (0, 2), (1, -9)\}$$

Do you see that the domain of the inverse of A is the same set as the range of A, and that the range of the inverse of A is the same as the domain of A? This is very important.

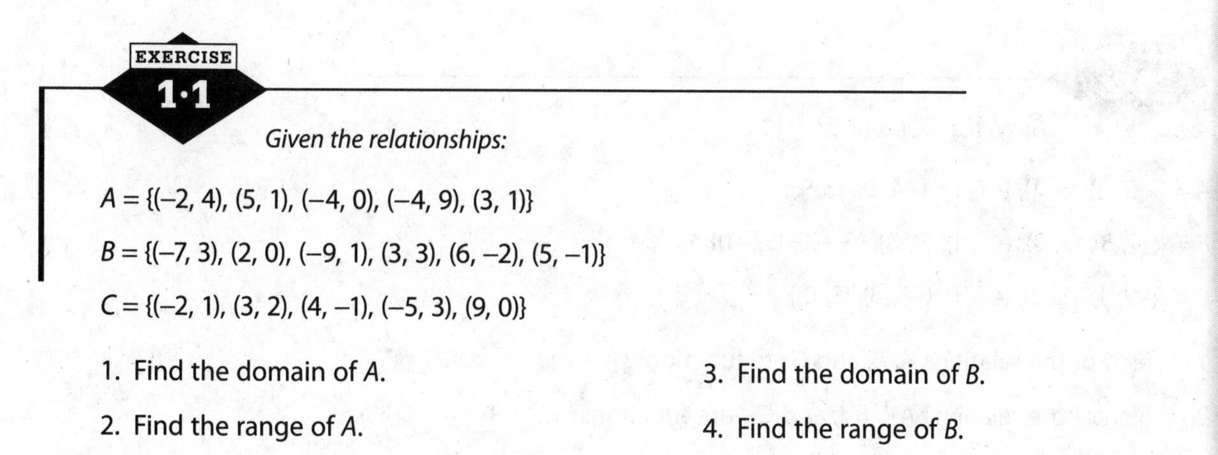

Given the relationships:

$A = \{(-2, 4), (5, 1), (-4, 0), (-4, 9), (3, 1)\}$

$B = \{(-7, 3), (2, 0), (-9, 1), (3, 3), (6, -2), (5, -1)\}$

$C = \{(-2, 1), (3, 2), (4, -1), (-5, 3), (9, 0)\}$

1. Find the domain of A.
2. Find the range of A.
3. Find the domain of B.
4. Find the range of B.

5. Find the domain of C.

6. Find the range of C.

7. Find A^{-1}.

8. Find B^{-1}.

9. Find C^{-1}.

Functions

Functions are a special case of a relation. By definition, a **function** is a relation in which each element of the domain (the input value) has a unique element in the range (the output value). In other words, for each input value there can be only one output value. Looking at the relations for A and A^{-1} in the previous section, you can see that A is not a function because the input value of 2 is associated with the output values 3 and 0. The relation A^{-1} is a function because each input value is paired with a unique output value. (Don't be confused that the number 2 is used as an output value for two different input values. The definition of a function does not place any stipulations on the output values.)

PROBLEM Given the relations:

$B = \{(-1, 3), (4, 7), (3, 2), (3, 5), (6, 3)\}$
$C = \{(0, -4), (12, 13), (-10, 7), (11, 9), (5, 13), (-4, 8)\}$

a. Determine if the relation represents a function.

b. Find the inverse of each relation.

c. Determine if the inverse of the relation is a function.

SOLUTION $B = \{(-1, 3), (4, 7), (3, 2), (3, 5), (6, 3)\}$

B is not a function because the input value 3 has two output values, 2 and 5.

$B^{-1} = \{(3, -1), (7, 4), (2, 3), (5, 3), (3, 6)\}$

B^{-1} is not a function because the input value 3 has two output values, –1 and 6.

$C = \{(0, -4), (12, 13), (-10, 7), (11, 9), (5, 13), (-4, 8)\}$

C is a function because each input value has a unique output value.

$C^{-1} = \{-4, 0), (13, 12), (7, -10), (9, 11), (13, 5), (8, -4)\}$

C^{-1} is not a function because the input value 13 has two output values, 12 and 5.

EXERCISE 1·2

Given the relationships:

$A = \{(-2, 4), (5, 1), (-4, 0), (-4, 9), (3, 1)\}$

$B = \{(-7, 3), (2, 0), (-9, 1), (3, 3), (6, -2), (5, -1)\}$

$C = \{(-2, 1), (3, 2), (4, -1), (-5, 3), (9, 0)\}$

1. Which of the relations A, B, and C are functions?

2. Which of the relations A^{-1}, B^{-1}, and C^{-1} are functions?

3. A relation is defined by the sets {(students in your math class), (telephone numbers at which they can be reached)}. That is, the input is the set of students in your math class and the output is the set of telephone numbers at which they can be reached. Must this relationship be a function? Explain.
4. Is the inverse of the relation in question 3 a function? Explain.
5. A relation is defined by the sets {(students in your math class), (the student's Social Security number)}. Must this relationship be a function? Explain.
6. Is the inverse of the relation in question 5 a function? Explain.
7. A relation is defined by the sets {(students in your math class), (the student's birthday)}. Must this relationship be a function? Explain.
8. Is the inverse of the relation in question 7 a function? Explain.

Function notation

Function notation is a very efficient way to represent multiple functions simultaneously while also indicating domain variables. Let's examine the function $f(x) = 5x + 3$. This reads as "f of x equals $5x + 3$." The name of this function is f, the independent variable (the input variable) is x, and the output values are computed based on the rule $5x + 3$. In the past, you would have most likely just written $y = 5x + 3$ and thought nothing of it. Given that, be patient as you work through this section.

What is the value of the output of f when the input is 4? In function notation, this would be written as $f(4) = 5(4) + 3 = 23$. Do you see that the x in the name of the function is replaced with a 4—the desired input value—and that the x in the rule of this function is also replaced with a 4? The point $(4, f(4))$ or $(4, 23)$ is a point on the graph of this function.

Consequently, you should think of the phrase $y =$ whenever you read $f(x)$. That is, if the function reads $f(x) = 5x + 3$ you should think $y = f(x) = 5x + 3$ so that you will associate the output of the function with the y-coordinate on the graph. Therefore (x, y) can also be written as $(x, f(x))$. $f(-2) = 5(-2) + 3 = -7$ indicates that when -2 is the input, -7 is the output, and the ordered pair $(-2, -7)$ is a point on the graph of this function.

Consider a different function, $g(x) = -3x^2 + 2x + 5$. $g(2) = -3(2)^2 + 2(2) + 5 = -12 + 4 + 5 = -3$. The point $(2, g(2))$ or $(2, -3)$ is on the graph of the parabola defined by $g(x)$. If $h(t) = -16t^2 + 128t + 10$, $h(3) = -16(3)^2 + 128(3) + 10 = -144 + 384 + 10 = 250$. Therefore $(3, h(3))$ represents the coordinate $(3, 250)$.

In essence, function notation is a substitution-guided process. Whatever you substitute within the parentheses on the left-hand side of the equation is also substituted for the variable on the right-hand side of the equation.

PROBLEM Given $f(x) = 4x^2 - 6x - 5$, find

a. $f(2)$

b. $f(0)$

c. $f(-3)$

SOLUTION a. $f(2) = 4(2)^2 - 6(2) - 5 = 16 - 12 - 5 = -1$

b. $f(0) = 4(0)^2 - 6(0) - 5 = 0 - 0 - 5 = -5$

c. $f(-3) = 4(-3)^2 - 6(-3) - 5 = 36 + 18 - 5 = 49$

PROBLEM Given $k(n) = 4n^2 + 3n - 2$, what does $k(2t - 1)$ equal?

SOLUTION Since $2t - 1$ is inside the parentheses, you are being told to substitute $2t - 1$ for n on the right-hand side of the equation.

$$\begin{aligned} k(2t-1) &= 4(2t-1)^2 + 3(2t-1) - 2 \\ &= 4(2t-1)(2t-1) + 3(2t-1) - 2 \\ &= 4(4t^2 - 4t + 1) + 6t - 3 - 2 \\ &= 16t^2 - 16t + 4 + 6t - 3 - 2 \\ &= 16t^2 - 10t - 1 \end{aligned}$$

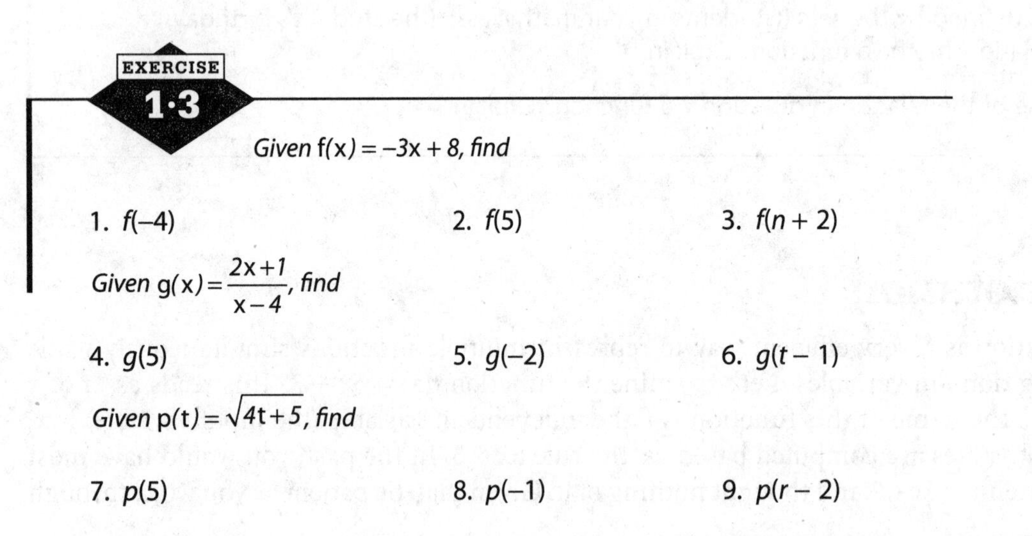

EXERCISE 1·3

Given $f(x) = -3x + 8$, *find*

1. $f(-4)$
2. $f(5)$
3. $f(n + 2)$

Given $g(x) = \dfrac{2x+1}{x-4}$, *find*

4. $g(5)$
5. $g(-2)$
6. $g(t - 1)$

Given $p(t) = \sqrt{4t+5}$, *find*

7. $p(5)$
8. $p(-1)$
9. $p(r - 2)$

Arithmetic of functions

Arithmetic can be performed on functions. For example, let $g(x) = 7x - 2$ and $p(x) = \dfrac{3x+2}{2x-3}$. To calculate $g(2) + p(2)$, you first evaluate each of the functions [$g(2) = 12$ and $p(2) = 8$] and then add the results: $g(2) + p(2) = 20$. $g(3) - p(1)$ shows that the input values do not have to be the same to do arithmetic. $g(3) = 19$ and $p(1) = -5$, so $g(3) - p(1) = 24$.

What does $p(g(2))$ equal? A better question to answer first is what does $p(g(2))$ mean? Since $g(2)$ is inside the parentheses for the function p, you are being told to make that substitution for x in the rule for p. It will be more efficient (and involve less writing) if you first determine that $g(2) = 12$ and evaluate $p(12)$. $p(12) = \dfrac{3(12)+2}{2(12)-3} = \dfrac{38}{21}$. Therefore, $p(g(2)) = \dfrac{38}{21}$. Evaluating a function with another function is called **composition of functions**. While $p(g(2)) = \dfrac{38}{21}$, $g(p(2)) = g(8) = 7(8) - 2 = 54$. This illustrates that you must evaluate a composition from the inside to the outside.

PROBLEM Given $f(x) = 4x^2 + 3$ and $g(x) = \dfrac{7x+1}{x+3}$, evaluate:

a. $f(3) + g(2)$
b. $g(-1) - f(-1)$
c. $f(g(2))$
d. $g(f(0))$

SOLUTION a. $f(3) = 4(3)^2 + 3 = 39$ and $g(2) = \frac{7(2)+1}{2+3} = \frac{15}{5} = 3$, so $f(3) + g(2) = 42$

b. $g(-1) = \frac{7(-1)+1}{-1+3} = \frac{-6}{2} = -3$ and $f(-1) = 4(-1)^2 + 3 = 4 + 3 = 7$, so $g(-1) - f(-1) = -10$.

c. $f(g(2)) = f(3) = 39$. (Look back at the solution to choice *a* for these answers.)

d. $g(f(0)) = g(3) = \frac{7(3)+1}{3+3} = \frac{22}{6} = \frac{11}{3}$.

As you know, there are two computational areas that will not result in a real number answer; thus, you do not (1) divide by zero, or (2) take the square root (or an even root) of a negative number. These rules are useful when trying to determine the domains of functions.

PROBLEM Find the domain for each function.

a. $f(x) = \frac{7x+1}{2x+3}$

b. $q(x) = \sqrt{3-6x}$

SOLUTION a. To avoid dividing by zero, $2x + 3 \neq 0$, so $x \neq -1.5$.

Therefore, the domain is all real numbers except -1.5. This is written as $\{x|x \neq -1.5\}$ and is read "all x such that $x \neq -1.5$."

b. $3 - 6x$ cannot be negative, so it must be the case that $3 - 6x \geq 0$ so $3 \geq 6x$ or $\frac{1}{2} \geq x$. Another way to write this is $x \leq \frac{1}{2}$.

The domain is read "all real numbers less than or equal to $\frac{1}{2}$."

Finding the range of a function is more challenging. This topic will be brought up throughout this book as particular types of functions are studied.

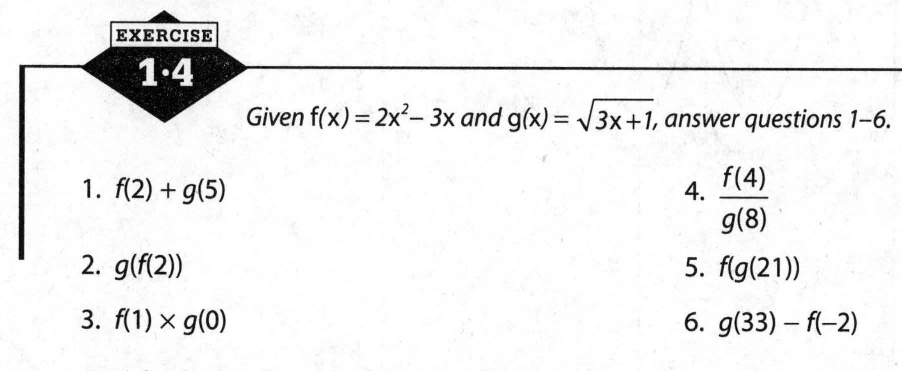

EXERCISE 1·4

Given $f(x) = 2x^2 - 3x$ *and* $g(x) = \sqrt{3x+1}$, *answer questions 1–6.*

1. $f(2) + g(5)$
2. $g(f(2))$
3. $f(1) \times g(0)$
4. $\frac{f(4)}{g(8)}$
5. $f(g(21))$
6. $g(33) - f(-2)$

Find the domain for each of the following functions.

7. $g(x) = \frac{2x+1}{x-4}$
8. $k(x) = \frac{2x+1}{5x^2-4x-1}$
9. $b(x) = \sqrt{4x+8}$
10. $a(x) = \sqrt{\frac{-1}{2}x+5}$

Transformation of functions

The graphs of $y = x^2$, $y = x^2 + 3$, $y = x^2 - 6$, $y = (x - 4)^2$, $y = (x + 5)^2$, and $y = (x + 1)^2 - 2$ are all parabolas. The difference among them is their location on the plane. Understanding the behavior of the base function, $y = x^2$, and the transformation that moves this function to a new location gives a great deal of information about the entire family of parabolas. Examine the following graphs.

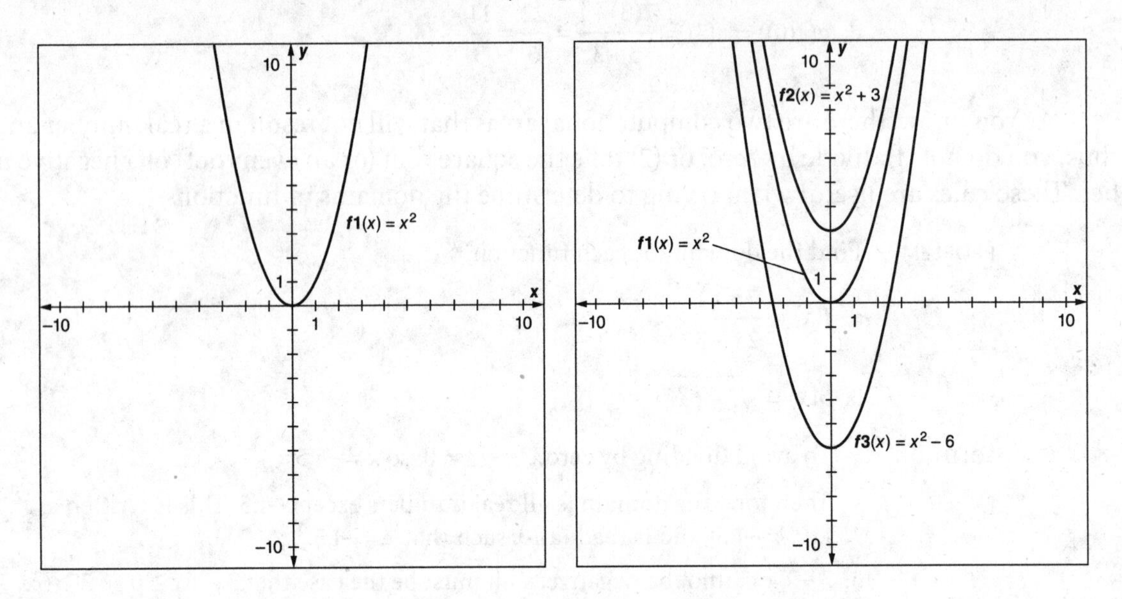

The graph of $y = x^2 + k$ is a vertical translation of the graph of $y = x^2$. If $k > 0$, the graph moves up, and if $k < 0$, the graph moves down. The graph of $y = (x - h)^2$ is a horizontal translation of the graph of $y = x^2$. The graph moves to the right when $h > 0$ and to the left when $h < 0$.

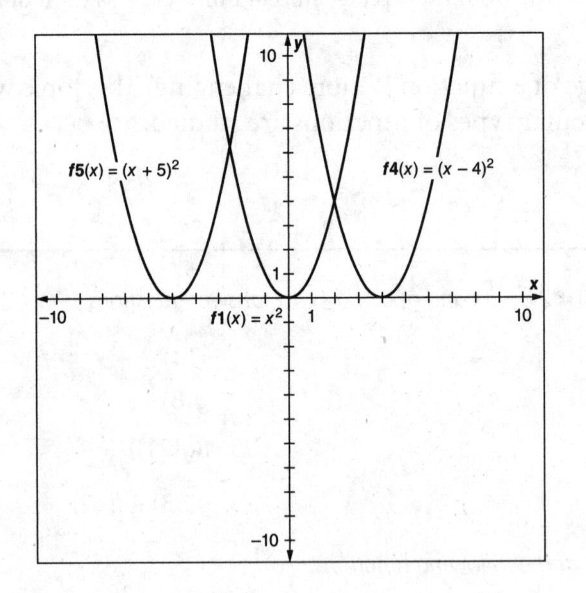

The transformation of $y = x^2$ to get the graph of $y = (x + 1)^2 - 2$ is a combination of the two. The graph of the parabola moves to the left 1 unit and down 2 units.

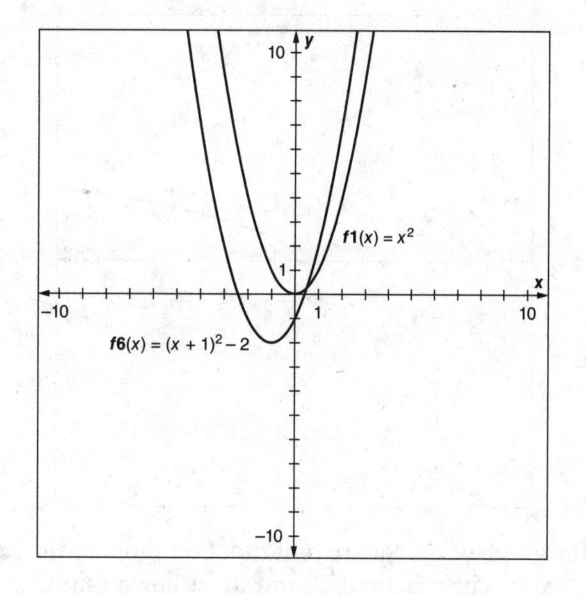

PROBLEM Describe the transformation for the function $g(x) = (x - 3)^2 + 1$.

SOLUTION The graph moves to the right 3 units and up 1 unit. (Use the graphing utility on your calculator to verify this.)

The graph of $y = ax^2$ is a stretch from the x-axis. It is important that you do not confuse the dilation from the origin that you studied in geometry (in which both the x- and y-coordinates are multiplied by the stretch factor) with a dilation from the x-axis (in which only the y-coordinate is multiplied by the stretch factor). If $0 < a < 1$, the graph moves closer to the x-axis, while if $a > 1$, the graph moves further from the x-axis. If $a < 0$, the graph is reflected over the x-axis.

PROBLEM Describe the transformation of $y = x^2$ to get the graph of $p(x) = -3(x + 2)^2 + 4$.

SOLUTION An easy way to do this is to follow what happens to an input value of x. The first thing that happens is 2 is added to the value of x (slide to the left 2 units), the value is squared (that is, the function in question), the result is multiplied by –3 (reflect over the x-axis and stretch from the x-axis by a value of 3), and 4 is added (slide up 4 units).

PROBLEM Describe the transformation of $y = x^2$ to get the graph of $q(x) = \frac{1}{2}(x - 4)^2 - 3$.

SOLUTION The graph of the parabola slides to the right 4 units, is stretched from the x-axis by a factor of $\frac{1}{2}$, and slides down 3 units.

PROBLEM The graph of the base function $y=\sqrt{x}$ is shown in the following diagram. Describe and sketch the graph of $f(x)=2\sqrt{x-3}-1$.

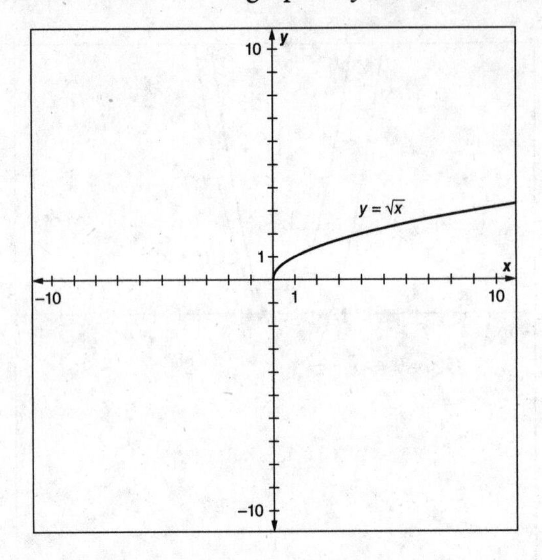

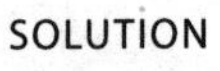

SOLUTION The graph of the square root function slides to the right 3 units, is stretched from the x-axis by a factor of 2, and slides down 1 unit.

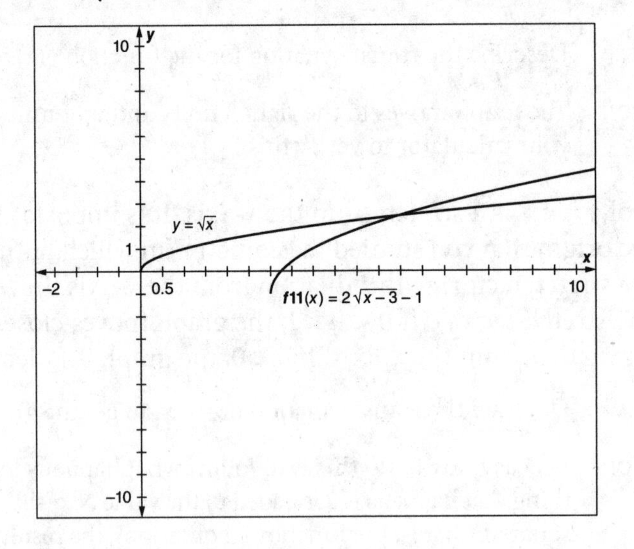

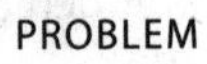

PROBLEM The graph of $y=|x|$ is shown in the following figure. Describe and graph the sketch of $y=-2|x+3|+1$.

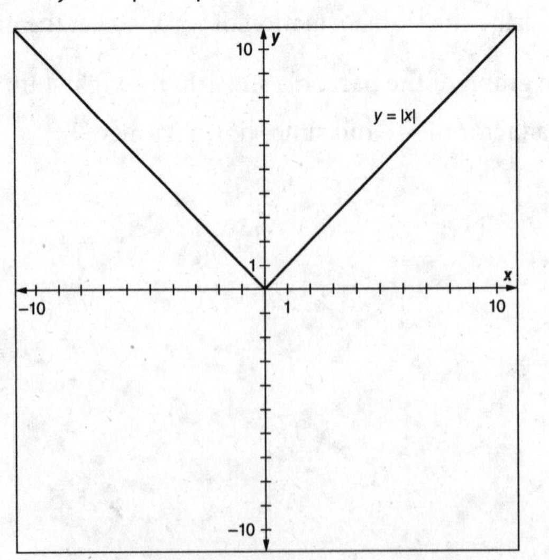

SOLUTION The graph of the absolute value function slides to the left 3 units, is reflected over the x-axis and stretched from the x-axis by a factor of 2, and moves up 1 unit.

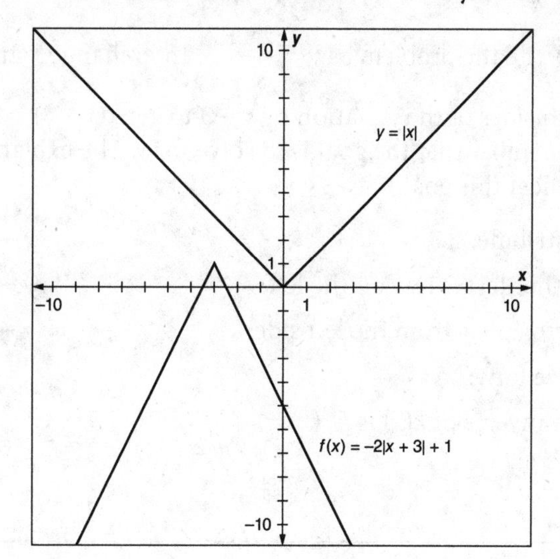

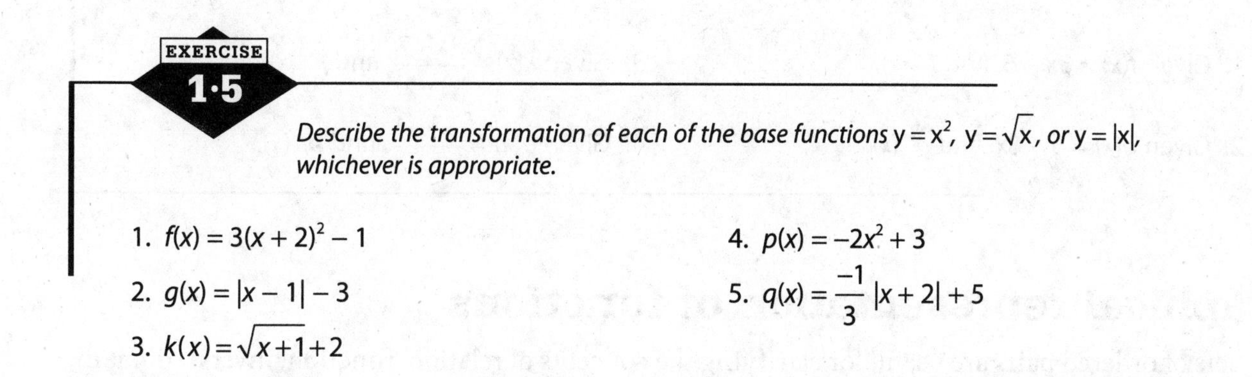

EXERCISE 1·5

Describe the transformation of each of the base functions $y = x^2$, $y = \sqrt{x}$, or $y = |x|$, whichever is appropriate.

1. $f(x) = 3(x + 2)^2 - 1$
2. $g(x) = |x - 1| - 3$
3. $k(x) = \sqrt{x+1} + 2$
4. $p(x) = -2x^2 + 3$
5. $q(x) = \dfrac{-1}{3}|x + 2| + 5$

Inverse of a function

To find the inverse of a function, the same notion of interchanging the x- and y-coordinates is applied. For example, to find the inverse of $f(x) = 5x + 3$, think about the function as $y = 5x + 3$. Switch the x and y: $x = 5y + 3$. Since functions are written in the form $y =$ rather than $x =$, solve the equation for y. Subtract 3 to get $x - 3 = 5y$ and then divide by 5 to get $y = \dfrac{x-3}{5}$. If $f(x) = 5x + 3$ then $f^{-1}(x) = \dfrac{x-3}{5}$.

PROBLEM Find the inverse function of $g(x) = \dfrac{2x+7}{3}$.

SOLUTION Rewrite the problem as $y = \dfrac{2x+7}{3}$. Interchange the x and y to get $x = \dfrac{2y+7}{3}$. Solve for y:

$3x = 2y + 7$ becomes $3x - 7 = 2y$ so $y = \dfrac{3x-7}{2}$. The inverse of $g(x)$ is $g^{-1}(x) = \dfrac{3x-7}{2}$.

PROBLEM Find the inverse of $k(x) = \frac{2x+7}{x-3}$.

SOLUTION Rewrite the problem as $y = \frac{2x+7}{x-3}$. Interchange x and y to get $x = \frac{2y+7}{y-3}$. Multiply both sides of the equation by $y - 3$ to get $x(y - 3) = 2y + 7$. (Important note: Remember that the goal is to solve for y. The explanation that follows is designed to meet this goal.)

Distribute: $xy - 3x = 2y + 7$

Gather terms in y on the left: $xy - 2y = 3x + 7$

Factor the y from the left side: $y(x - 2) = 3x + 7$

Solve for y: $y = \frac{3x+7}{x-2}$

The inverse of $k(x)$ is $k^{-1}(x) = \frac{3x+7}{x-2}$.

EXERCISE 1·6

For each function given, find the inverse.

1. Given $f(x) = 3x - 5$, find $f^{-1}(x)$.
2. Given $g(x) = 5 - 8x$, find $g^{-1}(x)$.
3. Given $k(x) = \frac{3x+7}{x-2}$, find $k^{-1}(x)$.
4. Given $p(x) = \frac{5x-1}{3x+4}$, find $p^{-1}(x)$.

Graphical representation of functions

Sets of ordered pairs are useful for clarifying the concepts of relation, function, inverse, domain, and range, but as you know, most of mathematics is done with formulas and graphs. By definition, a function is a relation in which no input value has multiple output values associated with it. What does that look like on a graph? The definition would indicate that it would not be possible to draw a vertical line anywhere on the graph and have it hit more than one of the plotted points at any one time. (If the vertical line does not hit any of the points, that is fine. The requirement is that the vertical line cannot hit more than one point at a time.)

PROBLEM Which of the following graphs represent functions?

a.

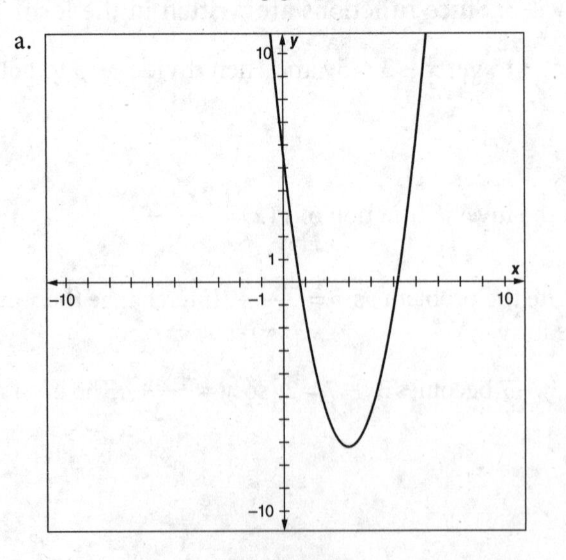

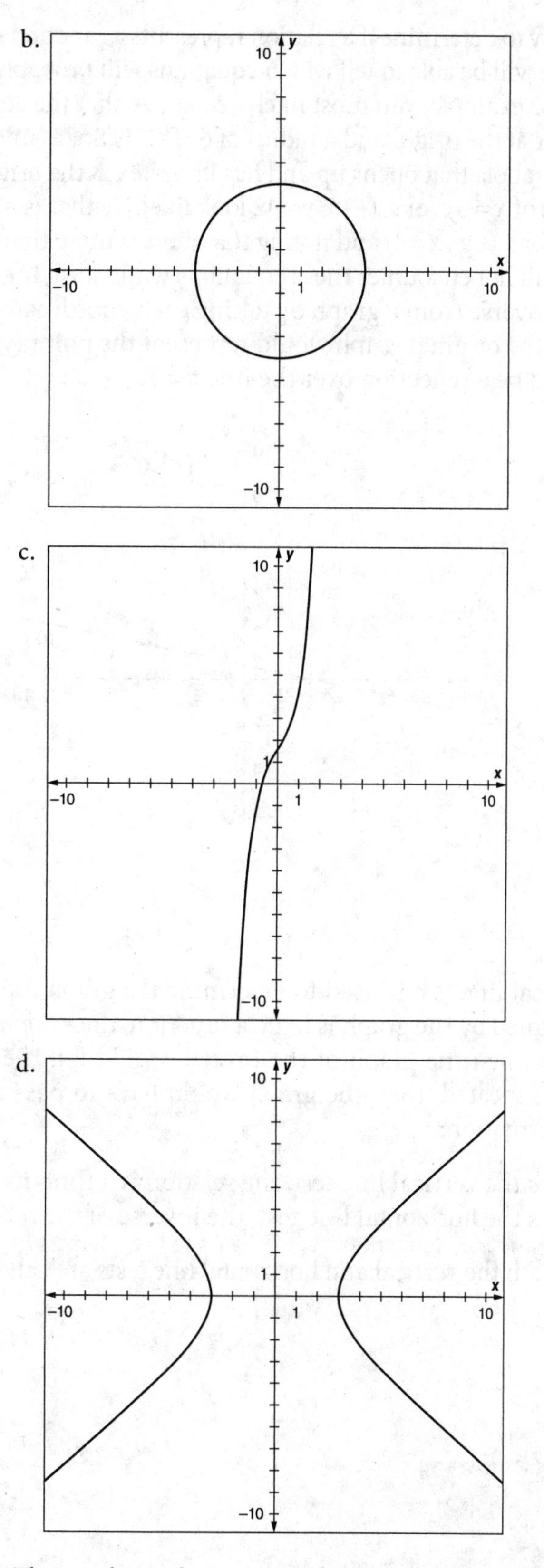

SOLUTION The graphs in choice *a* and choice *c* satisfy the vertical line test (i.e., a vertical line can never intersect either of these graphs at more than one point), while the remaining two graphs fail to satisfy the vertical line test (it is possible for a vertical line to intersect each of these graphs at more than one point).

At first, it is not as easy to determine if a relation represents a function when only given an equation. With experience, you will be able to tell which equations will probably not represent functions, and which are likely to. For example, you most likely recognize that the equation $x^2 + y^2 = 36$ represents a circle with its center at the origin and a radius of 6. This is not a function. You also know that the equation $y = 3x^2$ is a parabola that opens up and has its vertex at the origin. This is a function. Do you know what the graphs of $x = 3y^2$ or $xy^3 - x^3y = 12$ look like? Neither is a function, and this can be shown by picking a value for x (e.g., $x = 1$) and noting that there is more than one value of y associated with it. Fortunately, you will not encounter these equations while studying Algebra II.

You would find the inverse from a graph by taking each coordinate and interchanging the x and y values. So (x, y) on the original graph would represent the point (y, x) on the inverse graph. The resulting figure would be a reflection over the line $y = x$.

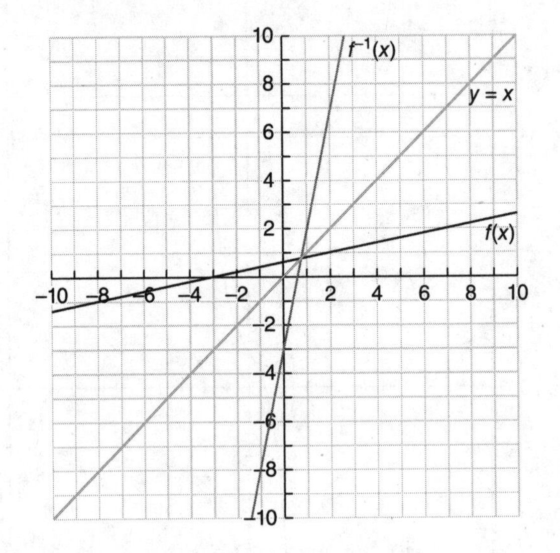

Recall that the vertical line test is used to determine if a graph represents a function. If the inverse of the relation defined by the graph is to be a function, then none of the y-coordinates can be repeated (if they were, then the graph of the inverse would fail the vertical line test). If the y-coordinates cannot be repeated, then the graph would have to pass a horizontal line test. To recap this important information:

- If a relation passes the vertical line test, the relation is a function.
- If a relation passes the horizontal line test, the inverse of the relation is a function.

Relations that pass both the vertical and horizontal line tests are called **one-to-one functions.**

PROBLEM Which of the following relations have inverses that are functions?

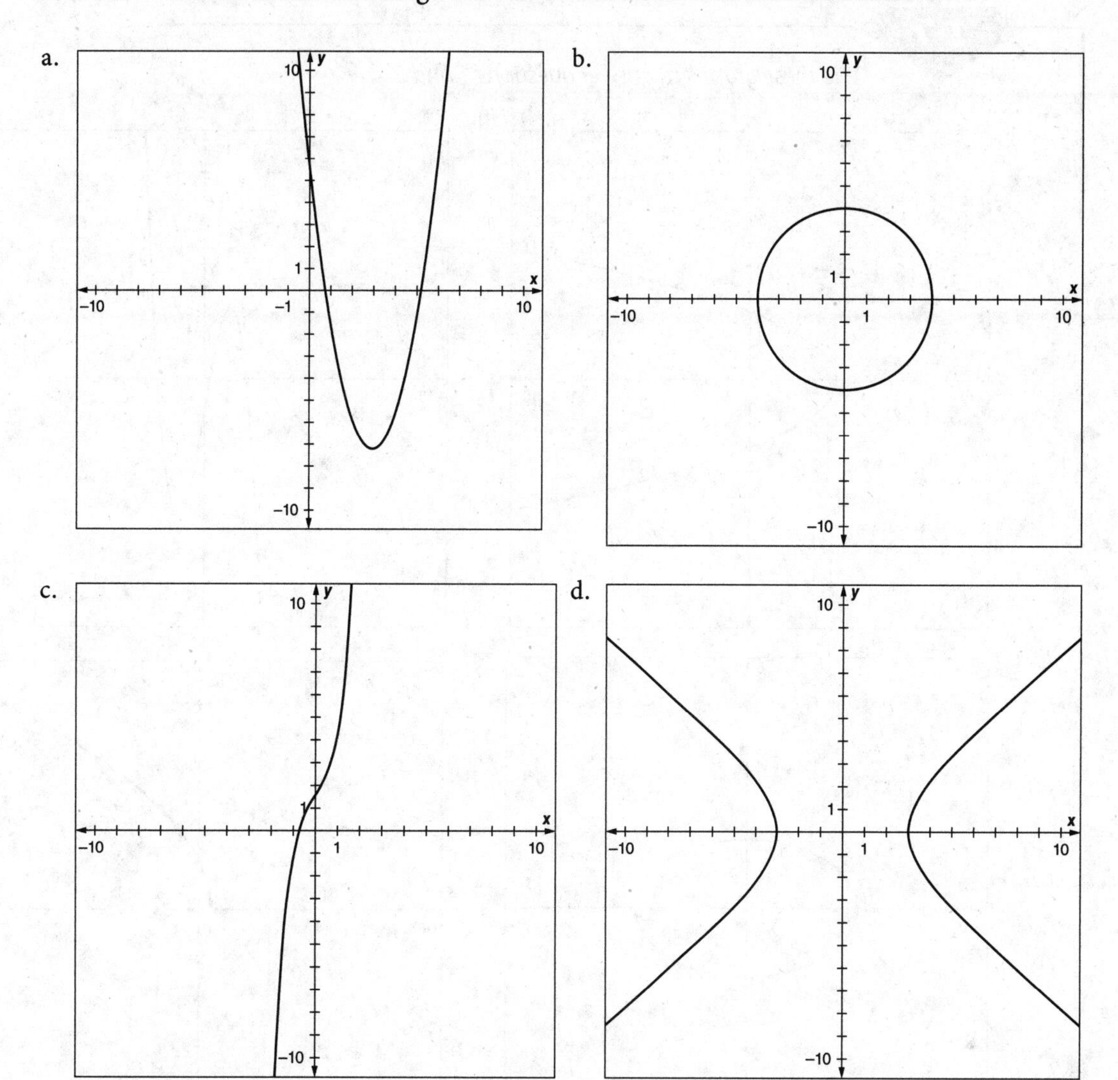

SOLUTION Only the graph in choice *c* passes the horizontal line test, so this is the only relation whose inverse will be a function.

EXERCISE
1·7

Use these graphs to answer questions 1 and 2.

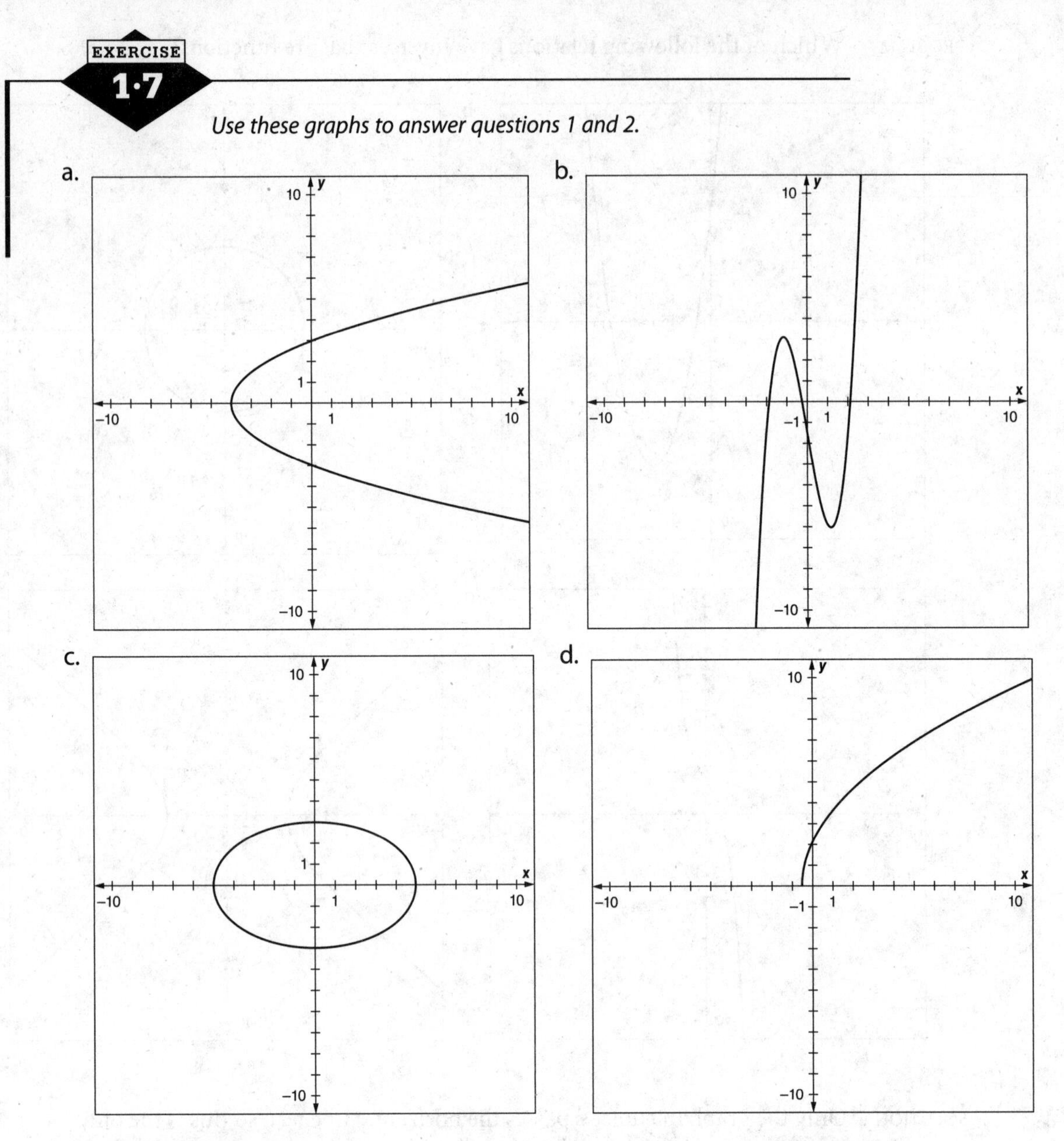

1. Which of the relations defined by the graphs given represent a function?
2. Which of the relations defined by the graphs given will have inverses that are functions?

Linear equations and inequalities

A constant theme in the study of mathematics is to take things back to basics. Linear equations and inequalities are the basic building blocks for the solution of all equations in mathematics.

Simple linear equations

All simple linear equations take the form $ax + b = c$. To find the solution, you would solve for x: $x = \frac{c-b}{a}$. The trick, of course, is to get the complicated "simple" linear equation into this basic form. The guiding principle is to gather common terms—those involving the variable in question on one side of the equation, and all other terms on the other side of the equation.

PROBLEM Solve $3(2x-5)-4(9-x)=29$.

SOLUTION The left-hand side of the equation requires the use of the distributive property twice. Within this, the minus sign preceding the second group of terms must not be ignored. $3(2x-5)=6x-15$, while $-4(9-x)=-36+4x$. Rewrite the equation to be

$$6x-15-36+4x=29$$

Simplify the left side: $10x-51=29$

Add 51: $10x=80$

Divide by 10: $x=8$

PROBLEM Solve $\frac{2x+7}{6}+\frac{3x-1}{10}=\frac{7}{15}$.

SOLUTION As most people would prefer to work with integers rather than fractions, this equation can be rewritten as an equivalent equation by multiplying both sides of the equation by a common denominator for 6, 10, and 15. The least common denominator (LCD) for these numbers is 30.

$$\left(\frac{2x+7}{6}+\frac{3x+1}{10}\right)30=\left(\frac{7}{15}\right)30$$

Distribute on the left: $\left(\frac{2x+7}{6}\right)30+\left(\frac{3x-1}{10}\right)30=\left(\frac{7}{15}\right)30$

Multiply fractions: $5(2x+7)+3(3x-1)=(7)2$

(Note: If you were expecting a different answer because of the direction to multiply, remember that part of the process of multiplying fractions is to simplify where possible.)

Distribute: $10x + 35 + 9x - 3 = 14$

Gather like terms: $19x + 32 = 14$

$19x = -18$

Solve: $x = \dfrac{-18}{19}$

Be warned: Not all answers will be "nice" integers.

PROBLEM Solve for x: $\dfrac{x+2b}{3a} - \dfrac{3x-4b}{6a} = \dfrac{c+5b}{a}$.

SOLUTION This is a **literal equation** because there is more than one letter in the problem. However, the directions indicate that x is the variable. All other letters in the problem will be treated as constants. The rules of combining terms will still be applied as the solution proceeds. The common denominator for these fractions is $6a$. Multiply both sides of the equation by this value.

$$\left(\frac{x+2b}{3a} - \frac{3x-4b}{6a}\right)6a = \left(\frac{c+5b}{a}\right)6a$$

Distribute: $\left(\dfrac{x+2b}{3a}\right)6a - \left(\dfrac{3x-4b}{6a}\right)6a = \left(\dfrac{c+5b}{a}\right)6a$

Multiply: $2(x + 2b) - (3x - 4b) = 6(c + 5b)$

Distribute: $2x + 4b - 3x + 4b = 6c + 30b$

Gather like terms: $-x + 8b = 6c + 30b$

$-x = 6c + 22b$

Divide by -1: $x = -22b - 6c$

There are many applications involving linear equations. Most involve systems of equations and will be looked at later in this chapter. The next two example problems are meant to highlight the importance of clearly defining the variable for an application and using the units of the problem to write an equation.

PROBLEM Colin's dad agreed to put any dimes or quarters he received as change into a piggy bank so that Colin could buy a new video game. They agreed that when there were 100 coins in the bank, Colin could have the money. When there were 100 coins in the bank, they discovered that the number of quarters in the bank was 11 less than twice the number of dimes. How many coins of each type were in the bank?

SOLUTION "How many coins of each type were in the bank?" gives a good hint as to how to define the variable—it is either the number of dimes or the number of quarters in the bank. "The number of quarters in the bank was 11 less than twice the number of dimes" relates the number of quarters in terms of the number of dimes. It will be easier to define the variable as the number of dimes in the bank.

Let d represent the number of dimes in the bank.

$2d - 11$

(11 less than twice the number of dimes) represents the number of quarters.

There were 100 coins in the bank: $d + 2d - 11 = 100$.

$$\begin{aligned} 3d &= 111 \\ d &= 37 \\ 2d - 11 &= 63 \end{aligned}$$

There were 37 dimes and 63 quarters in the bank.

Colin is off to a good start in saving for the new video game.

PROBLEM Colin's dad agreed to put any dimes or quarters he received as change into a piggy bank so that Colin could buy a new video game. They agreed that when there were 100 coins in the bank, Colin could have the money. When there were 100 coins in the bank, Colin determined that there was $19.45 in the bank. How many coins of each type were in the bank?

SOLUTION The question for this example is the same as the last example. However, the remaining information given in the problem is not about the number of coins, but the value of those coins. The equation for this problem needs to be about value, not quantity.

Let d represent the number of dimes in the bank.

The number of quarters must be the difference between 100 and the number of dimes, so the number of quarters is $100 - d$.

The equation is about money. Rather than using dollars as the units of the problem (and having to use decimals within the equation), the units that will be used in this problem will be cents:

$$\begin{aligned} 10d + 25(100 - d) &= 1945 \\ 10d + 2500 - 25d &= 1945 \\ -15d &= -555 \\ d &= 37 \\ 100 - d &= 63 \end{aligned}$$

There were 37 dimes and 63 quarters in the bank. Did you notice that the amount of money was written as 1945 (because the unit is cents) rather than 19.45? This is something that you will want to pay attention to as you work more complicated problems.

EXERCISE 2·1

Solve each of the following equations for the variable in the problem.

1. $4(2a - 3) - 3(8 - 2a) = -78$
2. $5y + 19 + 2(3 - 4y) = 7y - 12$
3. $\frac{3x-2}{8} + \frac{5x+9}{6} = \frac{17}{2}$
4. $\frac{8g+6}{12} - \frac{16g-3}{15} = \frac{19}{20}$
5. Solve for a: $3(2a - 5t) - 4(3a + 2t) = 8a - 25t$
6. Solve for q: $\frac{3q+2r}{4s} - \frac{4q-3r}{8s} = \frac{7r}{s}$

7. Kristen has categorized the songs on her MP3 player as rock (defined as any song her parents would not listen to) and classical (songs her parents would listen to). Kristen has 3670 songs on her MP3 player, and the number of rock songs is 110 more than seven times the number of classical songs. How many songs of each kind does she have on her MP3 player?

8. The garden department at Home Station is having a spring sale on plants. Diane bought a total of 90 plants for a total cost (before tax) of \$455. Diane only bought plants that were on sale either for \$4.50 each or for \$6.50 each. How many plants of each kind did she buy?

System of linear equations—graphical

Determining the values of the variables that make multiple equations true at the same time is important because most applications of mathematics involve the issue of meeting multiple requirements simultaneously. For example, business people want to know the point at which the money they spend to put products on the market—their cost—will be gained back from the money taken in by sales—their revenue. The point at which cost = revenue is called the **breakeven point.**

Systems of equations can be solved graphically as well as algebraically. In this section, you will study graphical solutions.

PROBLEM Sketch the graphs of $f(x) = 2x + 3$ and $g(x) = 5x - 9$ on the same set of axes.

What are the coordinates of the point at which the graphs intersect?

SOLUTION The point (4, 11) is the solution to this system of equations.

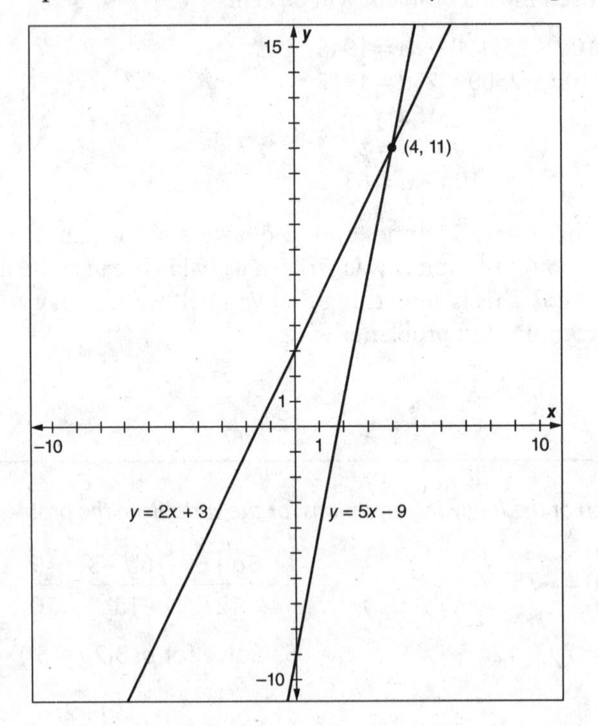

PROBLEM Sketch the graphs of $4x - 3y = 17$ and $y = 2x - 9$ on the same set of axes. What are the coordinates of the point at which the graphs intersect?

SOLUTION When using a graphing utility, equations need to be written in the form $y =$. Rewrite the equation $4x - 3y = 17$ to be $y = (4x - 17)/3$ in your equation editor. The point of intersection is (5, 1).

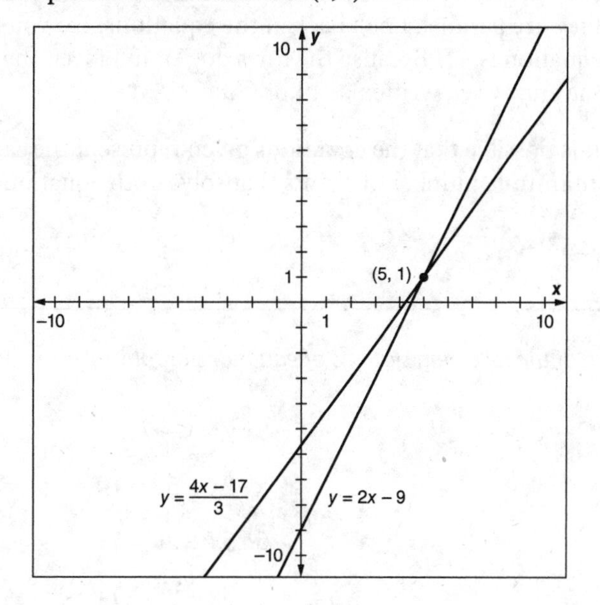

PROBLEM Sketch the graphs of $3x + 5y = -1$ and $4x - 3y = 18$ on the same set of axes. What are the coordinates of the point at which the graphs intersect?

SOLUTION When written in the form $y =$ for entry into the equation editor of your graphing utility, the equations become $y = \frac{-3x - 1}{5}$ and $y = \frac{4x - 18}{3}$. The graphs intersect at the point (3, –2).

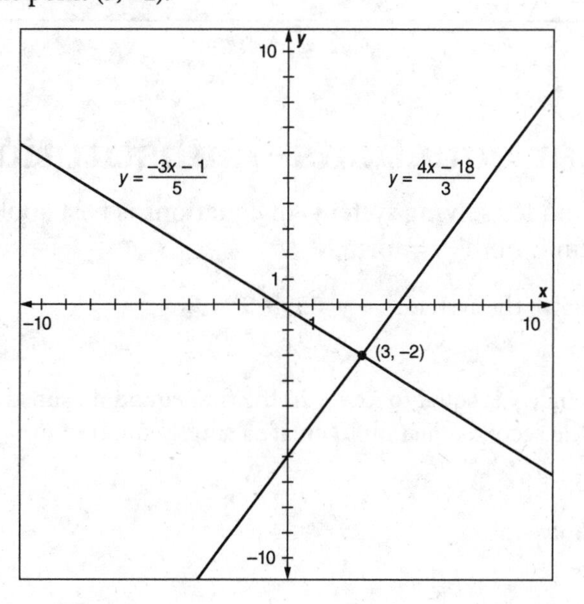

Please beware: When using graphing utilities, the window dimensions may need to be changed so that the point of intersection is visible on the screen.

PROBLEM Sketch the graphs of $4x + 2y = -3$ and $y = -2x + 8$ on the same set of axes. What are the coordinates of the point at which the graphs intersect?

SOLUTION When written in the form $y =$ for entry into the equation editor of your graphing utility, $4x + 2y = -3$ becomes $y = (-4x - 3)/2$. When viewed on the screen of your graphing utility, you can see that the two graphs will not intersect because they are parallel. Look back at the equations. Do you see that the slope for each equation is -2? Because the lines do not intersect, the solution to this system is the empty set, written as { } or ∅.

It is possible that the equations given represent the same line. In this case, there is an infinite number of points that solve both equations.

EXERCISE 2·2

Solve each of the following systems of equations graphically.

1. $y = x - 7$
   $y = 0.5x - 5$
2. $y = x + 6$
   $y = 2x + 11$
3. $y = x + 2$
   $5x + 3y = 42$
4. $y = 2x - 1$
   $3x - 2y = 4$
5. $y = x - 7$
   $2x + 5y = 14$
6. $5x + 2y = 11$
   $3x - y = -11$
7. $3x + 4y = 9$
   $-4x - 3y = 11$
8. $3x + 7y = 1$
   $8x + 19y = 1$

System of linear equations—substitution

The substitution method for solving systems of equations is best applied when at least one of the equations in the system is of the $y =$ form.

PROBLEM Solve the system: $y = 5x + 9$
$y = 3x - 5$

SOLUTION Since y is equal to $5x + 9$ in the first equation, substitute this expression for y in the second equation to create a single equation in x.

$5x + 9 = 3x - 5$

Solve for x: $2x = -14$
$x = -7$

Find the value of y: $y = 5x + 9$
$y = 5(-7) + 9 = -26$

The solution to this system is the ordered pair $(-7, -26)$.

PROBLEM Solve the system: $y = 8x + 15$
$5x - 7y = -3$

SOLUTION Substitute $8x + 15$ for y in the second equation to get:

$$5x - 7(8x + 15) = -3$$

Distribute: $5x - 56x - 105 = -3$

Gather like terms: $-51x = 102$

Solve for x: $x = -2$

Use the $y =$ equation to find y: $y = 8(-2) + 15 = -1$

The solution to this system is the ordered pair (–2, –1).

EXERCISE

2·3

Solve each system of equations using the substitution method.

1. $f(x) = x - 10$

   $g(x) = -3x + 2$

2. $f(x) = x + 7$

   $g(x) = 2x + 12$

3. $y = -x + 2$

   $2x + 3y = 16$

4. $y = x - 1.1$

   $2x + 7y = 49$

5. $y = x + 26$

   $-7x + 2y = 113$

System of linear equations—elimination

Although the substitution method can be used when the equations in the system are in standard form ($Ax + By = C$), the process is cumbersome and offers too many opportunities to make a mistake. The elimination (or multiplication–addition) method is a better choice. The goal in this method is to get the coefficients of one of the variables to be equal in size and opposite in sign.

PROBLEM Solve the system:

$3x - 2y = 23$
$8x + 3y = 3$

SOLUTION The coefficients of the y variables are already opposite in sign. Multiplying both sides of the first equation by 3 and both sides of the second equation by 2 will get these coefficients to have equal size.

$3(3x - 2y = 23)$
$\underline{2(8x + 3y = 3)}$
$9x - 6y = 69$
$16x + 6y = 6$

Add the two equations: $25x = 75$

Solve for x: $x = 3$

Solve for y:

$$8(3) + 3y = 3$$
$$24 + 3y = 3$$
$$3y = -21$$
$$y = -7$$

The solution to this system is the ordered pair (3, –7).

PROBLEM Solve the system:

$$5x + 4y = 115$$
$$7x + 6y = 167$$

SOLUTION The signs for all the coefficients are positive. Choosing to eliminate the y variable, multiply both sides of the first equation by 3 and both sides of the second equation by –2. (If you choose to, you could multiply the first equation by –3 and the second by 2.)

$$3(5x + 4y = 115)$$
$$\underline{-2(7x + 6y = 167)}$$
$$15x + 12y = 345$$
$$-14x - 12y = -334$$

Add the two equations: $x = 11$

Solve for y:

$$5(11) + 4y = 115$$
$$4y = 60$$
$$y = 15$$

The solution to this system of equations is the ordered pair (11, 15).

PROBLEM Solve the system:

$$\frac{2}{3}x + \frac{3}{5}y = 20$$
$$\frac{5}{6}x - \frac{3}{4}y = -5$$

SOLUTION While this equation can be solved by working with the coefficients as they are, you will be less likely to make a mistake if you first rewrite the equation so that the coefficients are integers rather than fractions. To that end, multiply both sides of the first equation by 15 (the common denominator for the fractions) and both sides of the second equation by 12.

$$10x + 9y = 300$$
$$10x - 9y = -60$$

Fortunately, these equations are ready to be added.

$$20x = 240$$
$$x = 12$$

Solve for y:

$$10(12) + 9y = 300$$
$$9y = 180$$
$$y = 20$$

Please note: You should always check your answer in the original equation to be sure you didn't make a mistake in an intermediate step of the solution.

EXERCISE

2·4

Solve each system of equations using the elimination method.

1. $2x - 5y = 29$

   $3x + 4y = 9$

2. $4x + 7y = 3$

   $-3x + 2y = 34$

3. $3x - 2y = 90$

   $4x - 3y = 75$

4. $12x + 9y = -4$

   $18x + 24y = 1$

5. $3x + 4y = 42$

   $7x - 9y = 21$

6. $5x + 3y = 25$

   $4x - 6y = -15$

System of linear equations—three variables

Equations in three variables can be graphed in a three-dimensional system—not something most classrooms have at their disposal. Equations in more than three variables do not have a physical representation available but they do represent the ability for the users of mathematics to think in abstract terms.

In this section, you will learn to solve systems of three linear equations in three variables using the elimination method. This method can be extended to any number of equations (having the same number of variables) to find a solution (if a solution exists). The process is to take one of the equations and pair it against the remaining equations. The same variable will be eliminated from each of these pairs, creating a new system of equations with one less equation and one less variable.

PROBLEM Solve the system:

$$\begin{aligned} 5x + 3y - 2z &= 51 \\ 2x - 4y + 3z &= -35 \\ 6x + 5y + 8z &= 19 \end{aligned}$$

SOLUTION Pair off the first equation with each of the other two.

$$\begin{aligned} 5x + 3y - 2z &= 51 \\ 2x - 4y + 3z &= -35 \end{aligned} \qquad \begin{aligned} 5x + 3y - 2z &= 51 \\ 6x + 5y + 8z &= 19 \end{aligned}$$

Remove z from both systems of equations.

$$\begin{array}{c} 3(5x + 3y - 2z = 51) \\ \underline{2(2x - 4y + 3z = -35)} \\ 15x + 9y - 6z = 153 \\ \underline{4x - 8y + 6z = -70} \\ 19x + y = 83 \end{array} \qquad \begin{array}{c} 4(5x + 3y - 2z = 51) \\ \underline{1(6x + 5y + 8z = 19)} \\ 20x + 12y - 8z = 204 \\ \underline{6x + 5y + 8z = 19} \\ 26x + 17y = 223 \end{array}$$

Add: $19x + y = 83$ $\quad 26x + 17y = 223$

Solve this new system of equations:

$$\begin{array}{r} 17(19x + y = 83) \\ \underline{-1(26x + 17y = 223)} \\ 323x - 17y = 1{,}411 \\ \underline{-26x - 17y = -223} \end{array}$$

Add: $297x = 1{,}188$

$$x = 4$$

Find y: $19(4) + y = 83$

$$y = 7$$

Find z: $5(4) + 3(7) - 2z = 51$

$$41 - 2z = 51$$

$$z = -5$$

The solution to this system of equations is the ordered triple (4, 7, –5).

PROBLEM Solve the system:

$$\begin{aligned} 24x - 30y + 36z &= 11 \\ 16x + 3y - 6z &= 9 \\ 4x + 12y - 6z &= 9 \end{aligned}$$

SOLUTION Pair off the third equation with each of the others.

$$\begin{aligned} 24x - 30y + 36z &= 11 \\ 4x + 12y - 6z &= 9 \end{aligned} \qquad \begin{aligned} 16x + 3y - 6z &= 9 \\ 4x + 12y - 6z &= 9 \end{aligned}$$

Eliminate z from each of the systems.

$$\begin{aligned} 24x - 30y + 36z &= 11 \\ \underline{6(4x + 12y - 6z = 9)} \\ 24x - 30y + 36z &= 11 \\ \underline{24x + 72y - 36z = 54} \end{aligned} \qquad \begin{aligned} 16x + 3y - 6z &= 9 \\ \underline{-1(4x + 12y - 6z = 9)} \\ 16x + 3y - 6z &= 9 \\ \underline{-4x - 12y + 6z = -9} \end{aligned}$$

Add: $48x + 42y = 65$ $\qquad 12x - 9y = 0$

This last equation can be simplified to $4x - 3y = 0$.

Solve this system:

$$\begin{aligned} 48x + 42y &= 65 \\ \underline{14(4x - 3y = 0)} \\ 48x + 42y &= 65 \\ \underline{56x - 42y = 0} \\ 104x &= 65 \\ x &= \frac{5}{8} \end{aligned}$$

Solve for y: $4\left(\frac{5}{8}\right) - 3y = 0$

$$3y = \frac{5}{2}$$

$$y = \frac{5}{6}$$

Solve for z: $4\left(\frac{5}{8}\right) + 12\left(\frac{5}{6}\right) - 6z = 9$

$$-6z = \frac{-7}{2}$$

$$z = \frac{7}{12}$$

The solution to this system of equations is the ordered triple $\left(\frac{5}{8}, \frac{5}{6}, \frac{7}{12}\right)$.

EXERCISE
2·5

Solve each system of equations using the elimination method.

1. $3x + 2y + 5z = 20$

   $4x - 3y - 2z = 9$

   $-2x - 5y + 8z = 43$

2. $5x - 3y - 2z = 27$

   $8x + 9y + 5z = 38$

   $11x + 3y - 3z = 73$

3. $4x + 5y - 3z = 34$

   $5x + 4y + 6z = 23$

   $8x - 9y - 5z = 66$

4. $7x - 6y + 6z = 17$

   $10x + 12y - 16z = 29$

   $11x - 9y - 2z = 35$

System of linear equations—matrix

In the traditional approach to solving the algebraic equation $ax = b$, you divide both sides of the equation by a to get the solution $x = \frac{b}{a}$. Because there is no operation called division in matrix algebra, the solution to the equation $AX = B$ is $X = A^{-1}B$, where A^{-1} is the inverse of matrix A. The matrix approach works best with a calculator that has matrix capabilities. While a matrix solution can be obtained using a pencil and paper approach, it will usually be much more cumbersome than the approaches shown in the earlier sections of this chapter.

The matrix equivalent to the system of equations

$$\begin{aligned} 5x + 4y &= 115 \\ 7x + 6y &= 167 \end{aligned}$$

has the coefficient matrix $A = \begin{bmatrix} 5 & 4 \\ 7 & 6 \end{bmatrix}$, variable matrix $X = \begin{bmatrix} x \\ y \end{bmatrix}$, and matrix of constants $B = \begin{bmatrix} 115 \\ 167 \end{bmatrix}$.

The solution is $X = A^{-1}B = \begin{bmatrix} 11 \\ 15 \end{bmatrix}$, or (11, 15) when written as an ordered pair.

The matrix solution to the system of equations

$$\begin{aligned} 5x + 3y - 2z &= 51 \\ 2x - 4y + 3z &= -35 \\ 6x + 5y + 8z &= 19 \end{aligned}$$

has coefficient matrix $A = \begin{bmatrix} 5 & 3 & -2 \\ 2 & -4 & 3 \\ 6 & 5 & 8 \end{bmatrix}$, variable matrix $X = \begin{bmatrix} x \\ y \\ z \end{bmatrix}$, and matrix of constants $B = \begin{bmatrix} 51 \\ -35 \\ 19 \end{bmatrix}$. The solution is $X = \begin{bmatrix} 4 \\ 7 \\ -5 \end{bmatrix}$, or the ordered triple (4, 7, −5).

PROBLEM Use matrices to solve the system of equations:

$$\begin{aligned} 5x + 4y - 2z &= 281 \\ 12x - 5y + 6z &= 513 \\ 8x - 3y - 3z &= -19 \end{aligned}$$

SOLUTION The coefficient matrix is $A = \begin{bmatrix} 5 & 4 & -2 \\ 12 & -5 & 6 \\ 8 & -3 & -3 \end{bmatrix}$, the variable matrix is $X = \begin{bmatrix} x \\ y \\ z \end{bmatrix}$, and the matrix of constants is $B = \begin{bmatrix} 281 \\ 513 \\ -19 \end{bmatrix}$. The solution to the system is $X = \begin{bmatrix} 37 \\ 51 \\ 54 \end{bmatrix}$.

A big advantage to solving systems of linear equations with a matrix solution is that the same amount of work is needed to solve a system with four equations in four variables as is needed to solve a system of equations with two equations in two variables.

PROBLEM Solve the system of equations:

$$\begin{aligned} 3w - x + 3y - 5z &= -316 \\ 5w + 2x + 8y - 5z &= 39 \\ 7w + 5x - 7y + 6z &= 659 \\ 11w - 7x - 5y + 17z &= 1454 \end{aligned}$$

SOLUTION The coefficient matrix is $A = \begin{bmatrix} 3 & -1 & 3 & -5 \\ 5 & 2 & 8 & -5 \\ 7 & 5 & -7 & 6 \\ 11 & -7 & -5 & 17 \end{bmatrix}$, the variable matrix is $X = \begin{bmatrix} w \\ x \\ y \\ z \end{bmatrix}$, and the matrix of constants is $B = \begin{bmatrix} -316 \\ 39 \\ 659 \\ 1454 \end{bmatrix}$. The solution is $X = \begin{bmatrix} 26 \\ 36 \\ 39 \\ 95 \end{bmatrix}$, or $w = 26$, $x = 36$, $y = 39$, and $z = 95$. The coordinates for the solution are the ordered 4-tuple (26, 36, 39, 95).

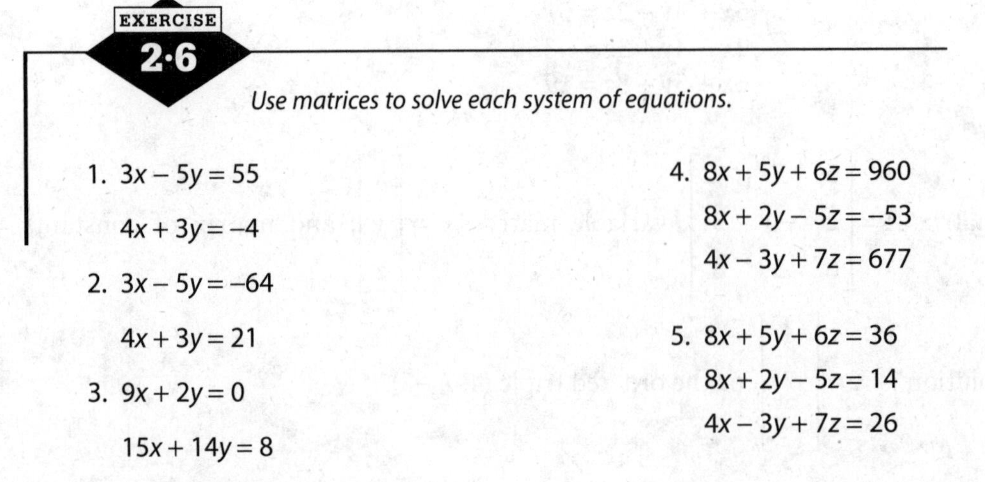

EXERCISE 2·6

Use matrices to solve each system of equations.

1. $3x - 5y = 55$
   $4x + 3y = -4$
2. $3x - 5y = -64$
   $4x + 3y = 21$
3. $9x + 2y = 0$
   $15x + 14y = 8$
4. $8x + 5y + 6z = 960$
   $8x + 2y - 5z = -53$
   $4x - 3y + 7z = 677$
5. $8x + 5y + 6z = 36$
   $8x + 2y - 5z = 14$
   $4x - 3y + 7z = 26$

6. $6x + 15y - 8z = 15$

   $18x - 25y + 16z = -7$

   $60x + 60y - 60z = 91$

7. $5w - 3x + 6y - 7z = 21$

   $4w + 3x - 2y + 8z = 37$

   $6w + 8x + 5y + 3z = 109$

   $w + 2x + 9y - z = 89$

8. $5w - 3x + 6y - 7z = 141$

   $4w + 3x - 2y + 8z = -23$

   $6w + 8x + 5y + 3z = 48$

   $w + 2x + 9y - z = 76$

System of linear equations—application

There are many applications of mathematics which can be solved with systems of linear equations. As you learned in Algebra I, it is to your advantage to clearly define the variables used to write the equations.

PROBLEM Diane asked Andrew to go shopping for candy for Colin's birthday party. Andrew knows that Colin prefers Nutters while some of his friends like Nougats. When he returned home, Andrew informed Diane that he bought a total of 30 candy bars and that he spent a total of \$20.70. Diane asked him how much each candy bar cost, and he responded that each Nutter cost \$0.75 and each Nougat cost \$0.60. After some thought, Diane said, "That should be enough of each. Thank you for going to the store." How many of each kind of candy bar did Andrew buy?

SOLUTION There are two types of information available in this problem: the number of candy bars purchased and the price (value) of each bar. Let t represent the number of Nutters purchased and g represent the number of Nougats purchased. The information can be displayed in two equations.

Number of candy bars: $t + g = 30$
Value of the candy bars (in cents): $75t + 60g = 2070$

Solve this system of equations using substitution, elimination, or matrices to determine that $t = 18$ and $g = 12$. Andrew bought 18 Nutters and 12 Nougats for the party.

PROBLEM Russ and Kate decided that they would put any loose change they accumulated each day into a jar, and then see how much was in the jar at the end of the month. When the month was over, Russ told Kate when she came home from work, "This is very interesting. We only put in nickels, dimes, and quarters. There was a total of \$63 in the jar from the 438 coins we put in." "Interesting," replied Kate. Russ went on, "Isn't it? It is also true that the amount of money in quarters we put in was twice the amount of money combined in dimes and nickels." "That means we put in..." started Kate, but was interrupted by a telephone call. How many of each kind of coin did they put into the jar?

SOLUTION There are two statements about the value of coins: the total value of coins and the relationship between the value of the quarters and the combined value of the nickels and dimes. There is also a statement about the total number of the coins.

Define the variables:

n represents the number of nickels in the jar
d represents the number of dimes in the jar
q represents the number of quarters in the jar

The statements about the number of coins can be represented by the equations:

Total number of coins: $n + d + q = 438$
Total value of the coins: $5n + 10d + 25q = 6300$
Quarters versus nickels and dimes: $25q = 2(5n + 10d)$

Rewrite the last equation to be $10n + 20d - 25q = 0$.

The system of equations is now:

$$\begin{aligned} n + d + q &= 438 \\ 5n + 10d + 25q &= 6300 \\ 10n + 20d - 25q &= 0 \end{aligned}$$

Use elimination or matrices to determine that $n = 120$, $d = 150$, and $q = 168$. Kate and Russ put in 120 nickels, 150 dimes, and 168 quarters.

PROBLEM The total resistance, R, for a parallel circuit composed of resistors with values R_1, R_2, R_3, etc. ohms (Ω) is given by the formula: $\frac{1}{R} = \frac{1}{R_1} + \frac{1}{R_2} + \frac{1}{R_3} + \cdots$ Three resistors with values R_1, R_2, and R_3 are used to create parallel circuits. If all three resistors are used, the total resistance of the circuit is $\frac{240}{23}\Omega$. If only R_1 and R_2 are used, the resistance in the circuit is 12Ω, and if only R_1 and R_3 are used, the resistance in the circuit is 16Ω. Determine the number of ohms in each of the three resistors, R_1, R_2, and R_3.

SOLUTION According to the formula for calculating the number of ohms in a parallel circuit, you get the equations:

$$\frac{1}{R_1} + \frac{1}{R_2} + \frac{1}{R_3} = \frac{23}{240}$$

$$\frac{1}{R_1} + \frac{1}{R_2} = \frac{3}{40}$$

$$\frac{1}{R_1} + \frac{1}{R_3} = \frac{1}{16}$$

Let x, y, and z represent $\frac{1}{R_1}$, $\frac{1}{R_2}$, and $\frac{1}{R_3}$, respectively. Use elimination or a matrix to determine $R_1 = 20\Omega$, $R_2 = 30\Omega$, and $R_3 = 80\Omega$.

PROBLEM In a modified version of a popular word game, words are made from letters on tiles. Some letters can earn double their point value when they land on a double letter space (indicated by underlining the letter). For example, if B is worth 5 points, U is worth 8 points, and D is worth 12 points, the word BUD is worth 25 $(5 + 8 + 12)$ points, and B<u>U</u>D is worth 33 $(5 + 2(8) + 12)$ points. Determine the value for each of the letters A, E, S, and T from these words: EA<u>S</u>T is worth 72 points, S<u>E</u>AT is worth 67 points, <u>T</u>EAS is worth 82 points, and S<u>A</u>TE is worth 64 points.

SOLUTION Using the variables e, a, s, and t, the problem becomes:

$\text{EA}\underline{\text{S}}\text{T}$	$e + a + 2s + t = 72$
$\text{S}\underline{\text{E}}\text{AT}$	$2e + a + s + t = 67$
$\underline{\text{T}}\text{EAS}$	$e + a + s + 2t = 82$
$\text{S}\underline{\text{A}}\text{TE}$	$e + 2a + s + t = 64$

Set up the matrix form for this problem to get:

$$\begin{bmatrix} 1 & 1 & 2 & 1 \\ 2 & 1 & 1 & 1 \\ 1 & 1 & 1 & 2 \\ 1 & 2 & 1 & 1 \end{bmatrix} \begin{bmatrix} e \\ a \\ s \\ t \end{bmatrix} = \begin{bmatrix} 72 \\ 67 \\ 82 \\ 64 \end{bmatrix}$$

Multiply by the inverse of the coefficient matrix:

$$\begin{bmatrix} e \\ a \\ s \\ t \end{bmatrix} = \begin{bmatrix} 1 & 1 & 2 & 1 \\ 2 & 1 & 1 & 1 \\ 1 & 1 & 1 & 2 \\ 1 & 2 & 1 & 1 \end{bmatrix}^{-1} \begin{bmatrix} 72 \\ 67 \\ 82 \\ 64 \end{bmatrix} = \begin{bmatrix} 10 \\ 7 \\ 15 \\ 25 \end{bmatrix}$$

E is worth 10 points, A is worth 7 points, S is worth 15 points, and T is worth 25 points.

Write systems of equations for each of the following problems. Solve the systems using the elimination method or a matrix equation.

1. Tickets for the fall drama production at Eastside High School were sold at three levels: student tickets purchased in advance, student tickets purchased on the day of the performance, and adult tickets (no matter when the tickets were purchased). There were three performances of the show: Friday night, Saturday night, and a Sunday matinee. The financial report shows the following results for ticket sales: Friday's show had 150 student advance tickets, 75 student tickets sold at the door, and 300 adult tickets; Saturday's show had 200 student advance tickets, 100 student tickets sold at the door, and 350 adult tickets; Sunday's show had 50 student advanced tickets, 70 student tickets sold at the door, and 250 adult tickets. Ticket receipts for the three nights were: Friday \$6300, Saturday \$7650, and Sunday \$4755. What was the price charged for each type of ticket?
2. Tickets for the spring musical at Bayview High School were being sold to students for \$8 in advance of the performance, \$10 on the day of the performance, and \$12 for adults (no matter when the ticket was purchased). The financial report after the musical performances showed that 1250 tickets were sold with receipts totaling \$13,400. The number of adult tickets sold exceeded the total number of student tickets sold by 150. How many tickets of each type were sold for the musical?
3. A newer version of the tiled word game has different values for each of the letters. Although $\underline{\text{underlined}}$ letters still score double their value, letters in **bold** font each triple their value. Determine the values for the letters B, A, E, and R if B$\underline{\text{A}}$RE is worth 111 points, **B**EA$\underline{\text{R}}$ is worth 113 points, **R**AB$\underline{\text{E}}$ is worth 97 points, and $\underline{\text{B}}$RA**E** is worth 99 points.

Linear inequalities

The most challenging thing to remember when solving simple linear inequalities is to reverse the orientation of the inequality when both sides of the sentence are multiplied or divided by a negative number.

PROBLEM Solve $4 - 3x > 13$. Graph the solution on a number line.

SOLUTION Subtract 4 to get $-3x > 9$, and then divide by -3 to get $x < -3$. (Notice the switch in the inequality.) Because this is a strict inequality ($<$, rather than $\leq$) an open circle is used to indicate the endpoint of the set.

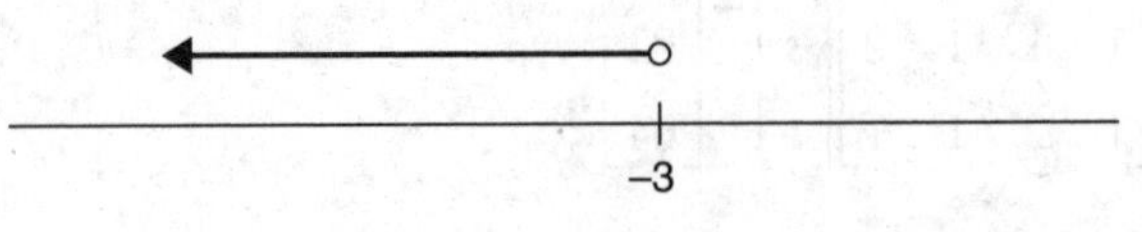

Examine the set of numbers graphed on the accompanying number line.

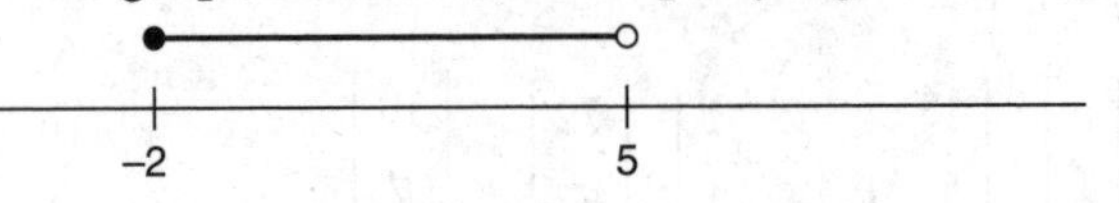

The set contains all the points from -2, which is included in the set, through 5, which is not included. That is, using x as the variable of the inequality, $x \geq -2$ and $x < 5$. This is usually written in the more condensed form $-2 \leq x < 5$ (x is between -2, included, and 5, excluded). This is an example of a compound inequality.

PROBLEM Solve $-9 < 4x + 3 \leq 17$ and graph the solution on a number line.

SOLUTION As discussed previously, $-9 < 4x + 3 \leq 17$ means $-9 < 4x + 3$ and $4x + 3 \leq 17$. The same steps will be used to solve both inequalities—subtract 3 and then divide by 4. Consequently, there is no need to split this compound inequality into its separate pieces.

$$-9 < 4x + 3 \leq 17$$

Subtract 3: $-12 < 4x \leq 14$

Divide by 4: $-3 < x \leq 3.5$

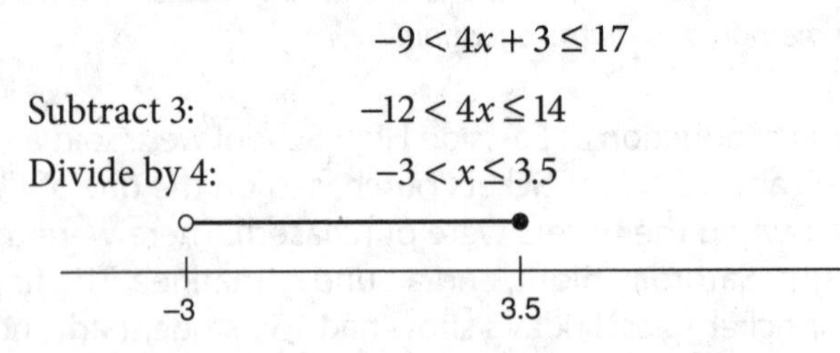

PROBLEM Solve $-7 < 5 - 2x \leq 11$ and graph the solution on a number line.

SOLUTION $-7 < 5 - 2x \leq 11$

Subtract 5: $-12 < -2x \leq 6$

Divide by -2: $6 > x \geq -3$

It is traditional to put the smaller number to the left when writing a compound inequality. The statement $-3 \leq x < 6$ is equivalent to the solution found.

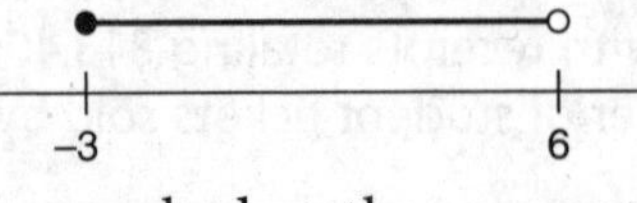

Examine the set of numbers graphed on the accompanying number line.

The set of numbers shows all those numbers that are less than or equal to –3 OR those numbers greater than 2. Written in mathematical notation, $x \leq -3$ or $x > 2$. It is important for you to realize that there is no other way to write this equality.

PROBLEM Solve $5x + 2 < 17$ or $3x - 9 > 12$. Graph the solution on a number line. Solve each of the inequalities separately.

SOLUTION

	$5x + 2 < 17$		$3x - 9 > 12$
Subtract 2:	$5x < 15$	Add 9:	$3x > 21$
Divide by 5:	$x < 3$	Divide by 3:	$x > 7$

The solution is $x < 3$ or $x > 7$.

PROBLEM Solve $5x + 2 > 17$ or $3x - 9 < 12$. Graph the solution on a number line.

SOLUTION This is similar to the last problem with the change being in the direction of the inequalities. The solution to this problem is $x > 3$ or $x < 7$. The graph of this solution makes it very clear that the solution to this problem is the set of real numbers, as every number on the number line is included in the solution.

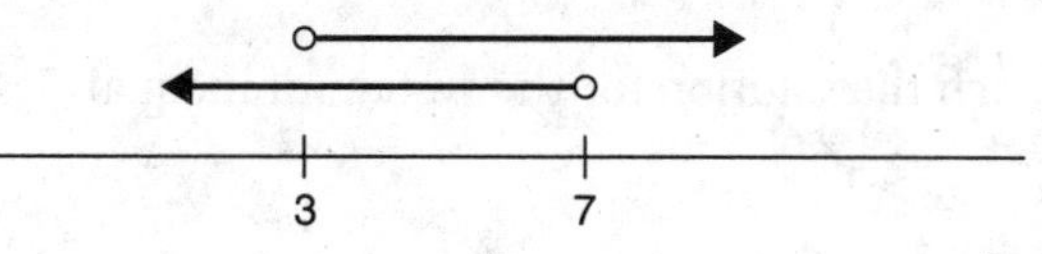

EXERCISE 2·8

Solve each of the following inequalities and graph the solution on a number line.

1. $8 + 3x > 5x - 4$
2. $2(3x - 5) - 3(7 - 4x) \geq 15x + 14$
3. $17 \leq 3x - 10 < 29$
4. $-8 < 12 - 5x \leq 17$
5. $3x - 2 < 12$ or $5x - 8 \geq 17$
6. $4 - 5x < 19$ or $3x + 10 \leq 25$

System of linear inequalities

When graphing inequalities on a number line, the difference in the graphs of $x > 1$ and $x \geq 1$ is to use an open circle at 1 for $x > 1$ (indicating that the endpoint is not included— this is called a **half line**) and a closed circle for $x \geq 1$ (indicating that the point is included—a **ray**). In the number plane, the graph of $x \geq 1$ would be a solid vertical line with the region to the right of the line shaded. The graph of $x > 1$ would also have the area to the right of the vertical line shaded, but the line would be a dotted line to show that the boundary is not included.

PROBLEM Sketch the common solution for the system of inequalities:

$y \leq 2x - 3$
$y > 4 - 3x$

SOLUTION The common solution is the overlapping region and is shaded a dark gray.

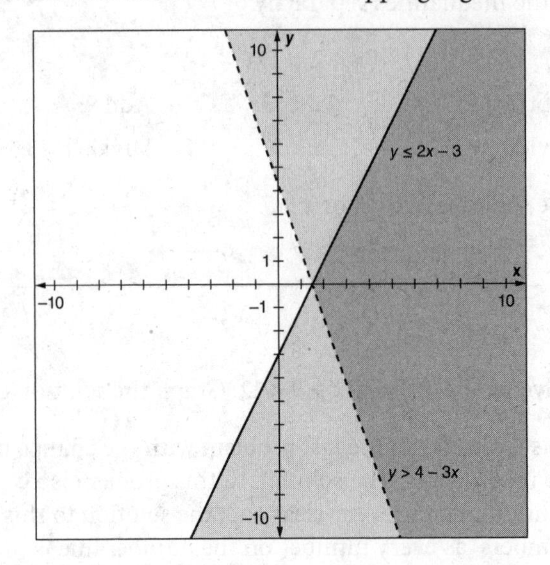

PROBLEM Sketch the solution for the system of inequalities:

$3x - 5y > 2$
$2x + y \leq 3$

SOLUTION Rewrite the first inequality as $y < \dfrac{3x - 2}{5}$ and enter that into your graphing utility.

The second inequality becomes $y \leq 3 - 2x$.

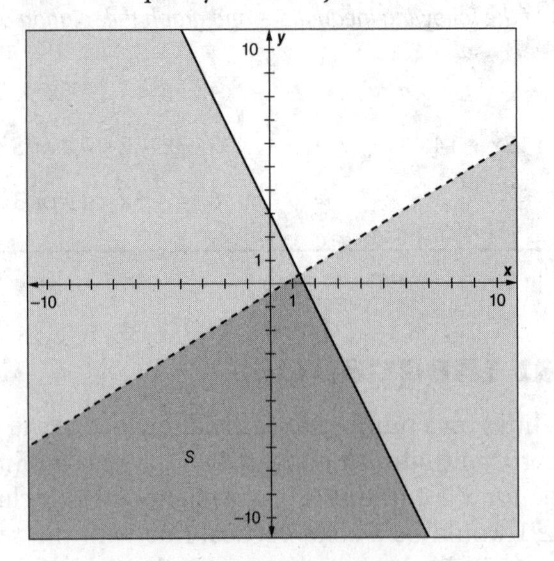

PROBLEM Sketch the solution for the system of inequalities:

$$y > -2$$
$$3x - y > 5$$
$$x + y < 2$$

SOLUTION All three boundaries are dotted, as there is no sense of equality in the open sentences. The screen can get crowded when graphing more than two inequalities, as shown in the next image.

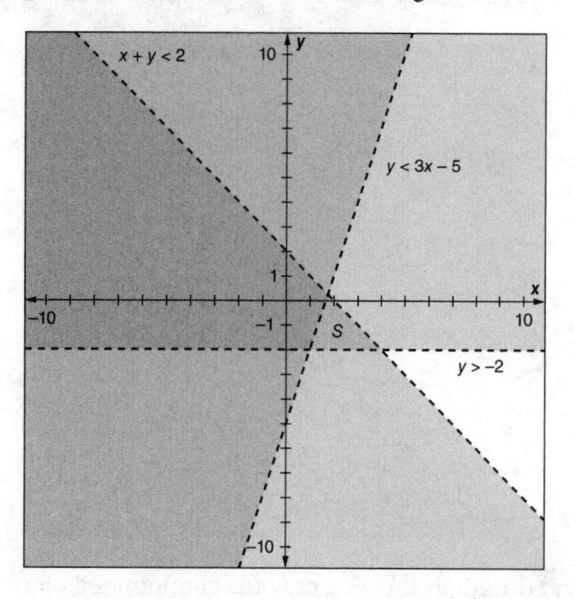

If only the common region is shaded, the common solution looks like this:

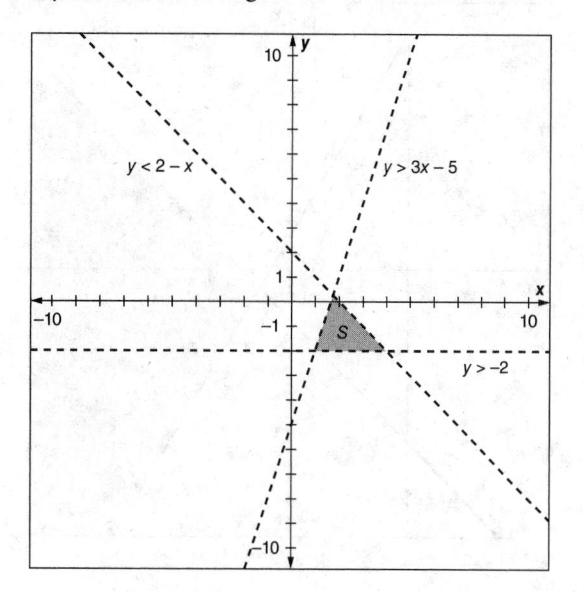

PROBLEM Sketch the solution for the system of inequalities:

$x > -5$
$x - y \geq 2$
$5x + 3y \geq 4$

SOLUTION Rewrite the last two inequalities as $y \leq x - 2$ and $y \geq \frac{4-5x}{3}$. The solution showing all three regions is:

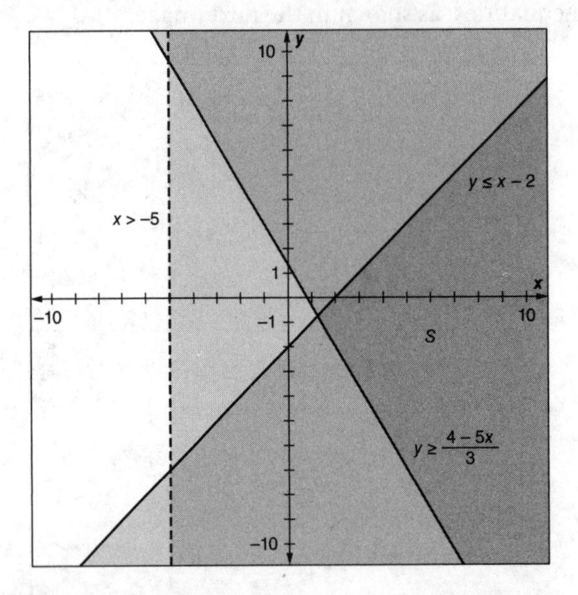

and the graph showing only the common region is:

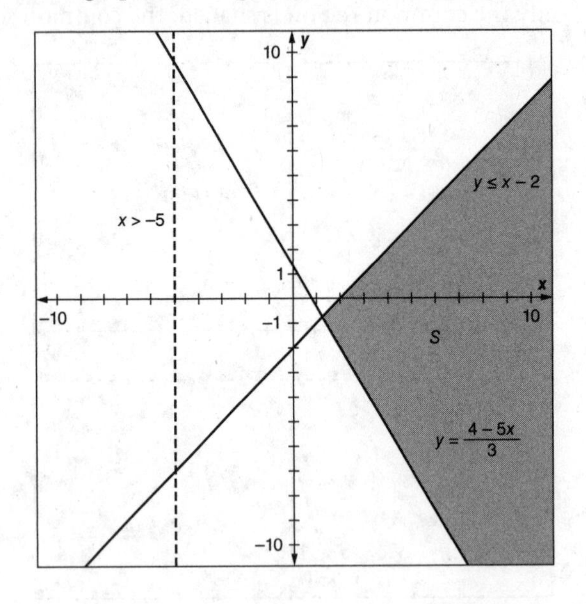

EXERCISE
2·9

Sketch the graphs for each of the systems of inequalities, showing only the common solution.

1. $x \geq 2$

   $y \leq 5$

   $y \geq x$

2. $x + y \leq 10$

   $2x - y < 3$

   $y < 7x$

3. $3x + 4y \leq 24$

   $4x + 3y > 12$

   $x < 5$

Absolute value equations

Most students learn the concept of **absolute value** as the magnitude of the number without the sign. For example, $|5| = 5$ and $|-5| = 5$. When these numbers (5 and −5) are graphed on a number line, you can see that both are 5 units from the origin. In fact, the geometric definition for absolute value is the distance a point is from the origin on the number line. This definition will prove helpful in solving absolute value equations and inequalities.

PROBLEM Solve $|x| = 5$.

SOLUTION As dicussed, $x = \pm 5$.

PROBLEM Solve $|x + 3| = 5$.

SOLUTION In this case, $x + 3 = \pm 5$. Solve $x + 3 = -5$ to get $x = -8$, and solve $x + 3 = 5$ to get $x = 2$. Observe that −8 and 2 are each 5 units from −3 and that the solution to $x + 3 = 0$ is $x = -3$. In other words, the solution to the equation $|x + 3| = 5$ is found by sliding the solution of $|x| = 5$ to the left 3 units.

PROBLEM Solve $|x - 4| = 3$.

SOLUTION Algebraic approach: $x - 4 = \pm 3$. Solve each of these equations to get $x = 1$ or $x = 7$.

Geometric approach: The solution to $x - 4 = 0$ is $x = 4$. The points 3 units from 4 on the number line are 1 and 7.

PROBLEM Solve $|2x - 5| = 3$.

SOLUTION Algebraic approach: $2x - 5 = \pm 3$. Solving each of these equations, $x = 1$ or $x = 4$.

Geometric approach: The coefficient of 2 has an impact on the solution. First, factor the 2 to get $2|x - 2.5| = 3$. Divide by 2 to get $|x - 2.5| = 1.5$. The solution to $x - 2.5 = 0$ is $x = 2.5$. Those points that are 1.5 units from 2.5 on the number line are 1 and 4.

PROBLEM Solve $|9 - 3x| = 12$.

SOLUTION Algebraic approach: $9 - 3x = \pm 12$. Solving each of these equations, $x = -1$ or 7.

Geometric approach: Factor -3 from $9 - 3x$ to get

$|-3(x-3)| = 12$.

Separate the factors: $|-3|\,|x-3| = 12$
Divide by $|-3|$: $|x-3| = 4$

Those points that are 4 units from 3 are -1 and 7.

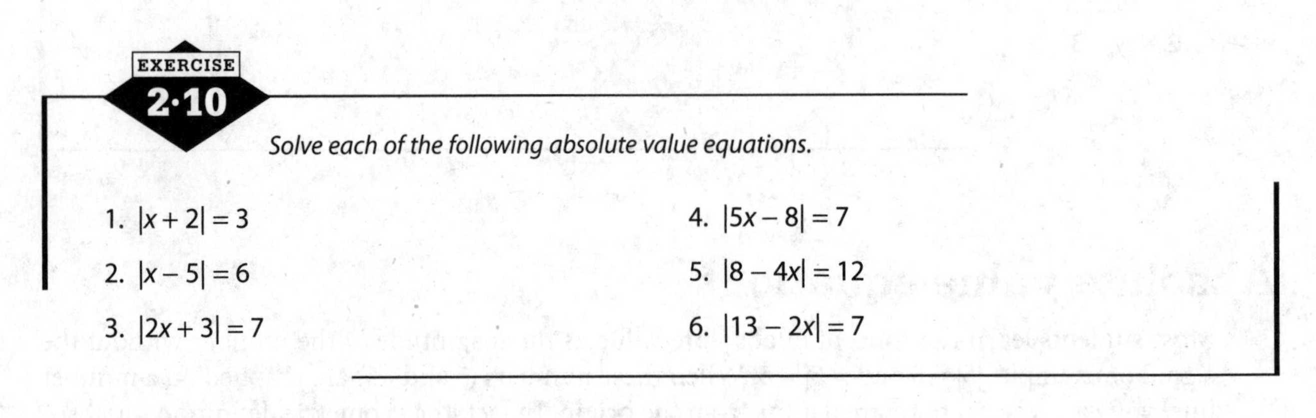

EXERCISE 2·10

Solve each of the following absolute value equations.

1. $|x+2| = 3$
2. $|x-5| = 6$
3. $|2x+3| = 7$
4. $|5x-8| = 7$
5. $|8-4x| = 12$
6. $|13-2x| = 7$

Absolute value inequalities

If $|x| = 5$ represents those points that are exactly 5 units from the origin on the number line, then it makes sense that $|x| > 5$ represents those points that are more than 5 units from the origin, and $|x| < 5$ represents those points that are less than 5 units from the origin. That is, the solution to $|x| > 5$ is $x > 5$ or $x < -5$, while the solution to $|x| < 5$ is $-5 < x < 5$.

PROBLEM Solve $|x-5| \le 3$ and graph the solution on a number line.

SOLUTION The absolute value of the expression is less than 3. Using the notion of distance from the preceding discussion, this translates to $-3 \le$ expression ≤ 3.

Substitute the expression: $-3 \le x - 5 \le 3$
Add 5: $2 \le x \le 8$

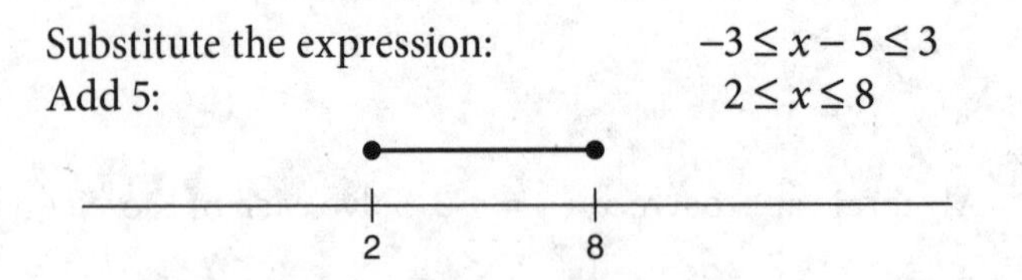

The point midway between 2 and 8 is 5. The graph shows all those numbers that are at most 3 units from 5.

PROBLEM Solve $|2x+5| \ge 7$ and graph the solution on a number line.

SOLUTION The absolute value of the expression is at least 7. The notion of distance indicates that the expression ≤ -7 or ≥ 7.

Substitute the expression: $2x + 5 \le -7$ or $2x + 5 \ge 7$
Subtract 5: $2x \le -12$ or $2x \ge 2$

Divide by 2: $x \le -6$ or $x \ge 1$

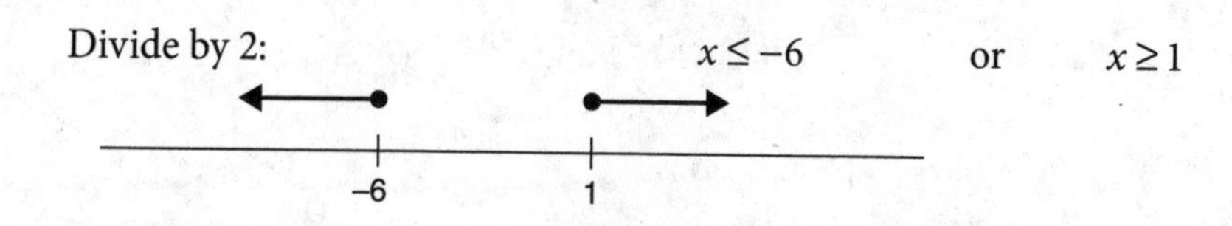

The point midway between -6 and 1 is -2.5 (the solution to $2x+5=0$). The points -6 and 1 are each 3.5 units from the point -2.5. Rewriting $|2x+5| \ge 7$ to $2|x+2.5| \ge 7$ and then $|x+2.5| \ge 3.5$ allows you to see the distance relationship.

PROBLEM Solve $|5-3x| > 8$ and graph the solution on a number line.

SOLUTION The absolute value of the expression is greater than 8. Therefore, the expression < -8 or > 8.

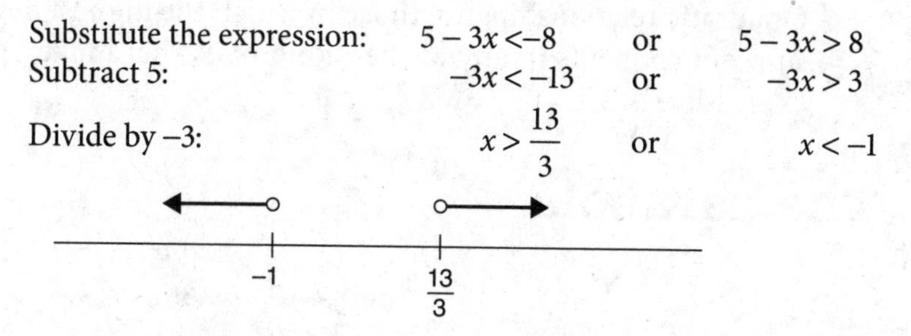

Substitute the expression:	$5-3x < -8$	or	$5-3x > 8$
Subtract 5:	$-3x < -13$	or	$-3x > 3$
Divide by -3:	$x > \frac{13}{3}$	or	$x < -1$

These points represent those points that are more than $\frac{8}{3}$ units from $\frac{5}{3}$.

PROBLEM Write an absolute value inequality for the numbers graphed on the number line.

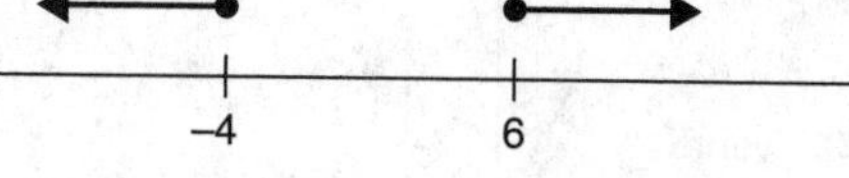

SOLUTION The point midway between -4 and 6 is 1. The points -4 and 6 are each 5 units from 1. The endpoints are included and the points represent values that are more than 5 units away. Therefore, an absolute value inequality for this graph is $|x-1| \ge 5$. (There are other absolute value inequalities that would also give this solution.)

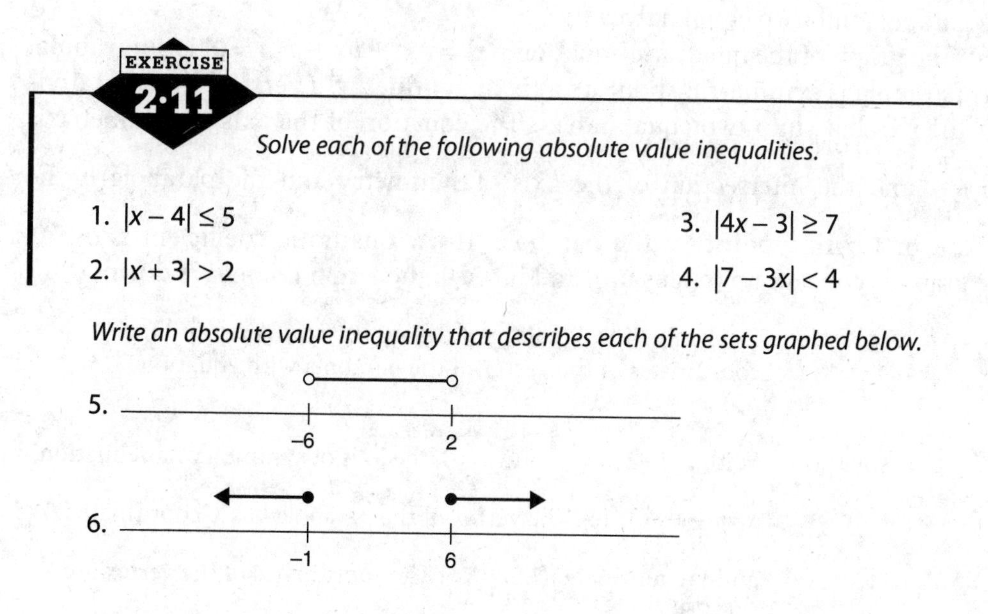

EXERCISE 2·11

Solve each of the following absolute value inequalities.

1. $|x-4| \le 5$
2. $|x+3| > 2$
3. $|4x-3| \ge 7$
4. $|7-3x| < 4$

Write an absolute value inequality that describes each of the sets graphed below.

5.

6.

Quadratic relationships

Quadratic relationships are those in which the highest degree (largest exponent or sum of exponents in any of the monomials that make up the polynomial) of the expression is 2.

The parabola

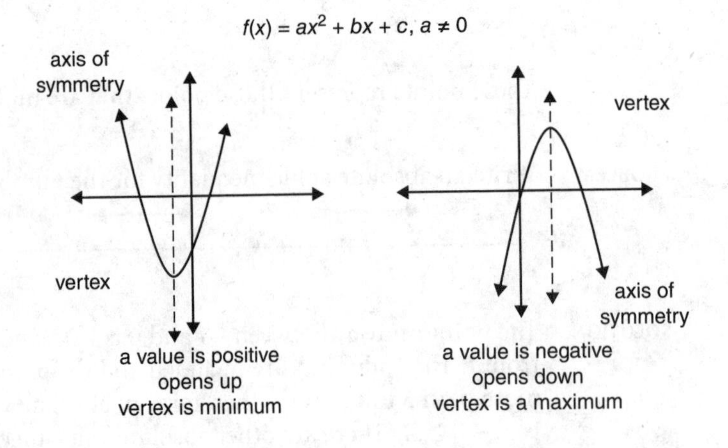

The parabola is symmetric about its axis of symmetry, a vertical line that divides the parabola into two equal halves.

The graph of the quadratic equation $f(x) = ax^2 + bx + c$ ($a \neq 0$) is a **parabola**. The parabola is symmetric about its **axis of symmetry**, a vertical line that divides the parabola into two equal halves. The equation of the axis of symmetry is $x = \frac{-b}{2a}$, and the intersection of the axis of symmetry and the parabola is the **vertex**, or **turning point**, of the parabola. If the quadratic coefficient is $a > 0$, the graph is concave up (opens up), and if $a < 0$, the graph is concave down.

PROBLEM Determine the equation of the axis of symmetry and the coordinates of the vertex of the parabola with equation $f(x) = -2x^2 + 6x + 3$.

SOLUTION With $a = -2$, $b = 6$, and $c = 3$, the axis of symmetry has equation $x = \frac{-6}{2(-2)} = 1.5$. The vertex of the parabola has x-coordinate 1.5 and y-coordinate $f(1.5) = 7.5$. The coordinates of the vertex are (1.5, 7.5).

If the parabola is concave up, the y-coordinate of the vertex represents the **minimum value** of the range of the function, and if the graph is concave down, the y-coordinate of the vertex represents the **maximum value** of the range of the function.

PROBLEM Determine the range of the function $f(x) = -2x^2 + 6x + 3$.

SOLUTION The quadratic coefficient is negative, so the graph is concave down. The coordinates of the vertex are (1.5, 7.5), so the range of the function is $y \leq 7.5$.

The ***x*-intercepts of a graph** correspond exactly to the roots of the equation $f(x) = 0$.

PROBLEM Find the x-intercepts of the function $g(x) = x^2 - 5x - 24$.

SOLUTION The x-intercepts can be found by solving the equation $x^2 - 5x - 24 = 0$. Factor the quadratic to get the equation $(x - 8)(x + 3) = 0$. The x-intercepts are (−3, 0) and (8, 0).

PROBLEM Write an equation for the parabola in the accompanying figure.

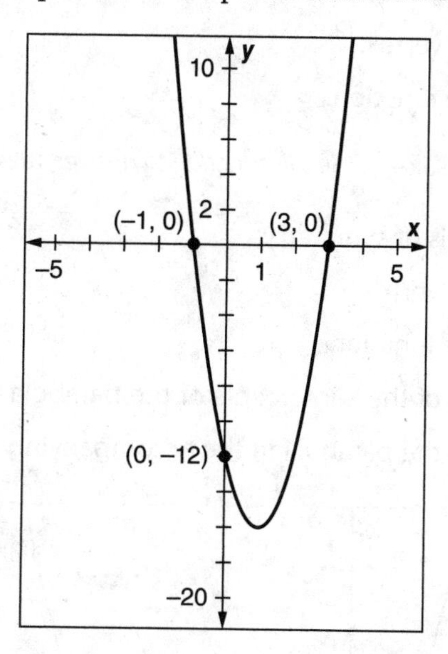

SOLUTION The y-intercept of this parabola is at −12, so $c = -12$ in the equation of the parabola. The x-intercepts are at −1 and 3. This means that the axis of symmetry must be the line midway between them, $x = 1$. Therefore, $1 = \frac{-b}{2a}$ or $2a = -b$. This can become $b = -2a$. Use the equation of the parabola and the x-intercept (−1, 0) to get the equation: $0 = a(-1)^2 + b(-1) - 12$

Substitute $-2a$ for b: $0 = a(-1)^2 + (-2a)(-1) - 12$

Solve for a: $12 = a + 2a = 3a$

$a = 4$

Solve for b: $b = -2a = -2(4) = -8$

The equation of the parabola is $f(x) = 4x^2 - 8x - 12$.

The form $f(x) = ax^2 + bx + c$ is called the **standard form** of the parabola. Another form of the parabola is called the **vertex form** and is written as $f(x) = a(x - h)^2 + k$. The vertex of this parabola is (h, k).

PROBLEM Determine the coordinates of the vertex of the parabola $f(x) = -3(x-5)^2 - 3$.

SOLUTION Using the vertex form, it can be determined that $x - h = x - 5$ so $h = 5$ and $k = -3$. The vertex of the parabola is at the point $(5, -3)$.

Because the axis of symmetry passes through the vertex of the parabola, it is also known that the axis of symmetry for $f(x) = -3(x-5)^2 - 3$ is $x = 5$.

Converting an equation from vertex form to standard form merely requires that the expression $a(x-h)^2 + k$ be expanded and like terms combined. Converting from standard form to vertex form requires a technique called completing the square that will be covered later in this chapter.

EXERCISE 3·1

Given the parabola $f(x) = 3x^2 + 12x - 5$. *Use this function to answer questions 1–3.*

1. Find the equation of the axis of symmetry.
2. Find the coordinates of the vertex.
3. Determine the range of the function.

Given the parabola $g(x) = \frac{-2}{3}x^2 + 6x + 4$, *use this function to answer questions 4–6.*

4. Find the equation of the axis of symmetry.
5. Find the coordinates of the vertex.
6. Determine the range of the function.
7. Determine the coordinates of the x-intercepts of the parabola $q(x) = x^2 - 11x + 30$.
8. Determine the equation of the parabola in the accompanying image.

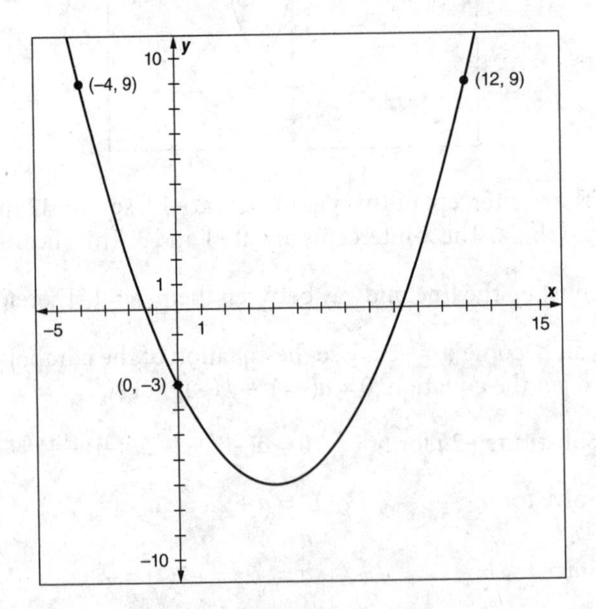

Given the parabola $p(x) = 2(x+3)^2 - 1$, *use this function to answer questions 9 and 10.*

9. Determine the coordinates of the vertex.
10. Write the equation for the axis of symmetry.

Special factoring formulas

The process for factoring is made easier if you know some basic, frequently used factoring patterns.

1. **Common factors:** The first step should always be to look for similarities among all the terms in the expression. The common factor may be a constant (as the 2 is in $2x^2 + 4x - 6$), a monomial (as $3x$ is in $6x^3 + 9x^2 - 12x$), or even a binomial (as $2x + 3$ is in $4x^2(2x + 3) - 11(2x + 3)$).

PROBLEM Factor: $2x^2 + 4x - 10$

SOLUTION $2x^2 + 4x - 10 = 2(x^2 + 2x - 5)$

PROBLEM Factor: $6x^3 + 9x^2 - 12x$

SOLUTION $6x^3 + 9x^2 - 12x = 3x(2x^2 + 3x - 4)$

PROBLEM Factor: $4x^2(2x + 3) - 11(2x + 3)$

SOLUTION $4x^2(2x + 3) - 11(2x + 3) = (2x + 3)(4x^2 - 11)$

2. **Difference of squares:** $a^2 - b^2 = (a + b)(a - b)$.

PROBLEM Factor: $16x^2 - 25y^2$

SOLUTION $16x^2 - 25y^2 = (4x)^2 - (5y)^2 = (4x + 5y)(4x - 5y)$

PROBLEM Factor: $64z^4 - 81p^6$

SOLUTION $64z^4 - 81p^6 = (8z^2)^2 - (9p^3)^2 = (8z^2 + 9p^3)(8z^2 - 9p^3)$

PROBLEM Factor: $75x^4 - 27x^2$

SOLUTION Neither 75 nor 27 is a square. However $3x^2$ is a common factor for each of the terms in the problem. Removing the common factor, $75x^4 - 27x^2$ becomes $3x(25x^2 - 9)$. The expression inside the parentheses meets the pattern of the difference of two squares. Completely factoring the original expression, $75x^4 - 27x^2 = 3x(5x + 3)(5x - 3)$.

3. **Square trinomials:** $a^2 + 2ab + b^2 = (a + b)^2$ and $a^2 - 2ab + b^2 = (a - b)^2$. Notice that the first and last terms are squares and that the middle term is twice the product of the square roots of the first and last terms. The sign of the middle term agrees with the sign inside the parentheses of the factored form.

PROBLEM Factor: $9x^2 - 30x + 25$

SOLUTION $9x^2 - 30x + 25 = (3x)^2 - 2(3x)(5) + (5)^2$; the first and last terms are squares and the middle term is twice the square root of the first and third terms. Therefore, $9x^2 - 30x + 25 = (3x - 5)^2$, with the sign between the binomial agreeing with the sign of the middle term of the original trinomial.

PROBLEM Factor: $36x^2 + 84xy + 49y^2$

SOLUTION $36x^2 + 84xy + 49y^2 = (6x)^2 + 2(6x)(7y) + (7y)^2 = (6x + 7y)^2$

PROBLEM Factor: $(2x + 3)^2 + 2(2x + 3)(3x - 5) + (3x - 5)^2$

SOLUTION As difficult as this looks at first, recognize that this is the pattern for the square trinomial, and $(2x + 3)^2 + 2(2x + 3)(3x - 5) + (3x - 5)^2 = (2x + 3 + 3x - 5)^2 = (5x - 2)^2$.

4. **Difference and sum of cubes:** $a^3 - b^3 = (a - b)(a^2 + ab + b^2)$ and $a^3 + b^3 = (a + b)(a^2 - ab + b^2)$. You need to be careful with these formulas because they look very similar to the square trinomials. Make note of the differences: the sign between the terms in the binomial is the same as the sign between the original cubes, and the sign of the middle term in the factor is the opposite of this; the coefficient of the middle term in the trinomial factor is 1 (not 2, as it is in the square trinomial terms).

PROBLEM Factor: $27x^3 - 125$

SOLUTION $27x^3 - 125 = (3x)^3 - (5)^3 = (3x - 5)((3x)^2 + (3x)(5) + (5)^2) = (3x - 5)(9x^2 + 15x + 25)$

PROBLEM Factor: $\frac{1}{8}x^6 + 64$

SOLUTION $\frac{1}{8}x^6 + 64 = \left(\frac{1}{2}x^2\right)^3 + (4)^3 = \left(\frac{1}{2}x^2 + 4\right)\left(\left(\frac{1}{2}x^2\right)^2 - \left(\frac{1}{2}x^2\right)(4) + (4)^2\right) =$

$\left(\frac{1}{2}x^2 + 4\right)\left(\frac{1}{4}x^4 - 2x^2 + 16\right)$

EXERCISE 3·2

Completely factor each of the following.

1. $24x^2 - 16x$
2. $15x^2 + 240$
3. $8x^4 - 12x^3 + 20x^2$
4. $8x^2(4x - 3) + 3x(4x - 3) - 9(4x - 3)$
5. $x^2 - y^2$
6. $49y^2 - 25$
7. $81t^2 - 144$
8. $(3x + 4)^2 - (2y - 1)^2$
9. $a^2 + 18ab + 81b^2$
10. $100x^2 + 140xy + 49y^2$
11. $49a^2 - 42ab + 9b^2$
12. $100w^2 - 100wv + 25v^2$
13. $x^3 + 125$
14. $1 - 64x^3$
15. $16x^3 + 54$
16. $216 - 125k^3$

Trial and error

Perhaps a good way to introduce the notion of the trial and error method of factoring is to look at two multiplication problems.

PROBLEM Multiply: $(x + 9)(x - 8)$

SOLUTION

Use the distributive property:	$(x + 9)(x) + (x + 9)(-8)$
Distribute a second time:	$(x)(x) + (9)(x) + (x)(-8) + (9)(-8)$
Multiply pairs:	$x^2 + 9x - 8x - 72$
Combine like terms:	$x^2 + x - 72$

You should recognize that the x^2 is a result of multiplying the first terms in the parentheses, the -72 came from multiplying the last terms, and the middle term x came from adding 9 and -8.

PROBLEM Multiply: $(4x+9)(15x-8)$

SOLUTION Use the distributive property: $(4x+9)(15x)+(4x+9)(-8)$

Distribute a second time: $(4x)(15x)+(9)(15x)+(4x)(-8)+(9)(-8)$

Multiply pairs: $60x^2+135x-32x-72$

Combine like terms: $60x^2+103x-72$

The $60x^2$ came from multiplying the first terms in the parentheses, the -72 came from multiplying the last terms, and the middle term $103x$ came from the sum of the product of the first term in the first factor and the second term in the second factor, and the product of the second term in the first factor and the first term in the second factor. (Read that again to make sure you have it.)

Being asked to factor $60x^2+103x-72$ without having first seen the multiplication done might be a fairly daunting experience for many people. 60 has a number of factors (1 and 60, 2 and 30, 3 and 20, 4 and 15, 5 and 12, 6 and 10), as does 72 (1 and 72, 2 and 36, 3 and 24, 4 and 18, 6 and 12, 8 and 9). Determining which pairs of these factors will give the correct middle term can be a challenge.

There are some basic arithmetic facts that can assist in the solution of this kind of problem. There are also some approaches that use technology, one of which will be examined in the next section.

Arithmetic approach

The simple concepts that a negative product comes from the multiplication of numbers with different signs, and that an odd sum comes from adding an even and an odd number, can help eliminate some of the multitude of factors that can arise in a problem like $60x^2+103x-72$. The third term constant is negative, so you know that the factors of 72 shown in the parentheses in the previous solution will have opposite signs, and that 103 came from the difference rather than the sum of terms. Because the middle term is odd, there must be an odd product and an even product from the multiplication of the inner and outer terms of the original binomials. As a result, 2 and 30, and 6 and 10 can be eliminated as candidates for the factors of 60, and 2 and 36, 4 and 18, and 6 and 12 can be eliminated as factors of 72. The original problem has no common factors so neither of the binomial factors can have a common factor. 3 and 24 can be eliminated as candidates for 72. $72-60$ does not equal 103 so the factors 1 and 60 and 1 and 72 can be eliminated from consideration. This leaves 3 and 20, 4 and 15, and 5 and 12 as the possible factors of 60; and 8 and 9 as the factors of 72. Check (the trial aspect) the various products of the factors of 60 with 8 and 9. $(9)(15)=135$ and $(4)(8)=32$ and the difference between them is 103. Therefore, the factors of $60x^2+103x-72$ are $4x+9$ and $15x-8$.

Graphical approach

Use a graphing utility to sketch the graph of $f(x)=60x^2+103x-72$.

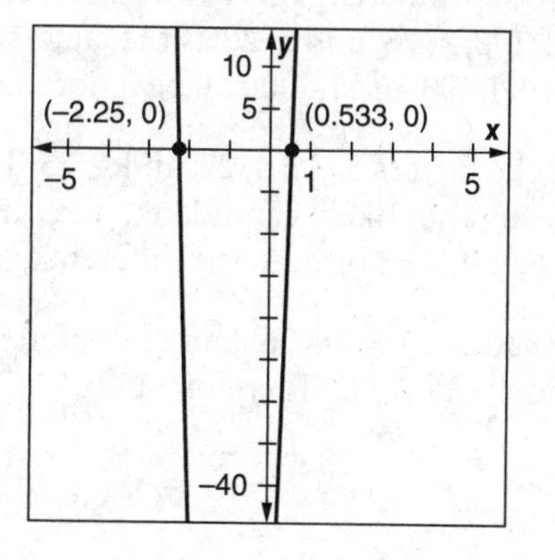

Since it is only the x-intercepts that need to be seen, the viewing window does not need to be set to see the vertex of the parabola.

Use the zero feature of your calculator to find the x-intercepts of the graph. The image below was created using a TI-Nspire. The intercepts are associated with the variables **x1** and **x2**. Use the approxRational command to convert these numbers to fractions. The command parameters allow for a tolerance level, and in these cases, 0.005 is chosen.

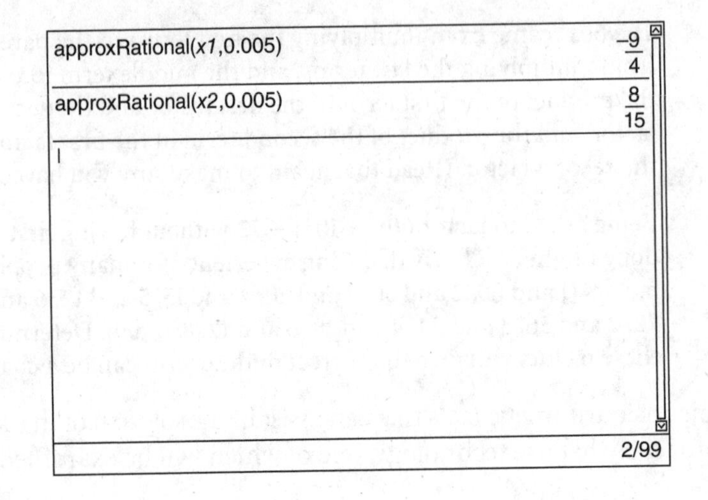

The zeroes of the function are $x = \frac{-9}{4}$ and $x = \frac{8}{15}$. To get the factors of the original quadratic, manipulate each of these equations back to binomials.

Multiply by the denominators:	$4x = -9$	$15x = 8$
Set one side to zero:	$4x + 9 = 0$	$15x - 8 = 0$

When factored, $60x^2 + 103x - 72 = (4x + 9)(15x - 8)$.

PROBLEM Factor $108x^2 + 177x - 100$

SOLUTION Algebraic approach: The factors of 108 and 100 are

108: 1, 108; 2, 54; 3, 36; 4, 27; 6, 18; 9, 12

100: 1, 100; 2, 50; 4, 25; 5, 20; 10, 10

The last term is negative so the signs need to be opposite. The middle term is odd so the products of the outside and inside terms of the binomial factors must have one odd result and one even result. Therefore, factors of 108 and 100 that are both even can be eliminated. This leaves 1, 108; 3, 36; 4, 27; and 9, 12 as the factors for 108, and 1, 100; 4, 25; and 5, 20 as the factors for 100. 108 – 100 does not equal 177 so the pairs 1, 108 and 1, 100 can be ignored.

Consider the factors 3, 36; 4, 27; and 4, 25. The products 25×36 and 25×27 are very large and not likely candidates. The difference between 4×36 and 3×25 does not equal 177, nor does the difference between 4×27 and 4×25.

$25 \times 12 = 300$ but $4 \times 9 = 36$, and the difference is not 177. The difference between 25×9 and 4×12 is 177. Therefore, $108x^2 + 177x - 100$ factors to $(12x + 25)(9x - 4)$.

Graphical approach: The following graph shows $f(x) = 108x^2 + 177x - 100$ with zeroes.

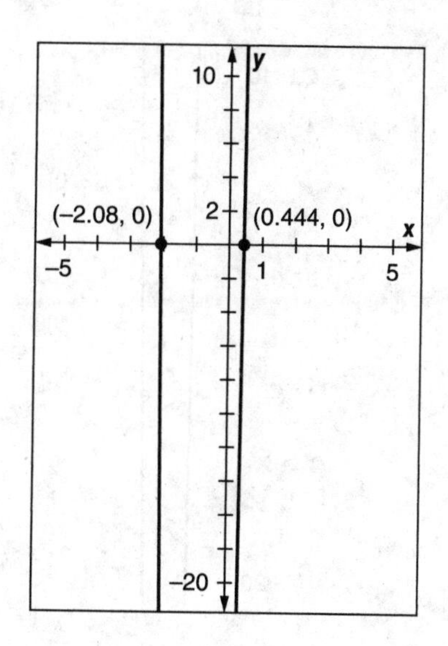

Convert each of the zeroes to fractions to get $x = \frac{-25}{12}$ and $x = \frac{4}{9}$. Rewrite these as binomials: $12x + 25 = 0$ and $9x - 4 = 0$. Therefore, $108x^2 + 177x - 100$ factors to $(12x + 25)(9x - 4)$.

PROBLEM Factor: $112x^2 - 414x + 315$

SOLUTION Algebraic approach: The third term is positive so the signs of the factors are the same, and the sign of the middle term is negative, so both factors are negative. 414 is the sum of the products of the outside and inside terms of the binomial factors and these products are either both odd or both even. The factors of 112 and 315 are:

112: 1, 112; 2, 56; 4, 28; 7, 16; 8, 14

315: 1, 315; 3, 105; 5, 63; 7, 45; 9, 35; 15, 21

All of the factor pairs for 315 are odd. 1, 112 and 7, 16 are factors of 112 with one odd and one even and will not provide the appropriate middle term. 414 is a fairly large number, but the products of 105 with any of the larger factors of 112 will give a number much larger than 414. Will the product of 105 with the smaller factors of 112 work? The only factor pairs to consider are $105 \times 2 = 210$ and $56 \times 3 = 168$. The sum of these numbers is not 414. 105 multiplied by the other factors gives a result greater than 414. $15 \times 8 = 120$ and $21 \times 14 = 294$ and the sum of these numbers is the required 414. Therefore, $112x^2 - 414x + 315$ factors to $(8x - 21)(14x - 15)$.

Graphical approach: The graph of $f(x) = 112x^2 - 414x + 315$ and the zeroes are shown in the image.

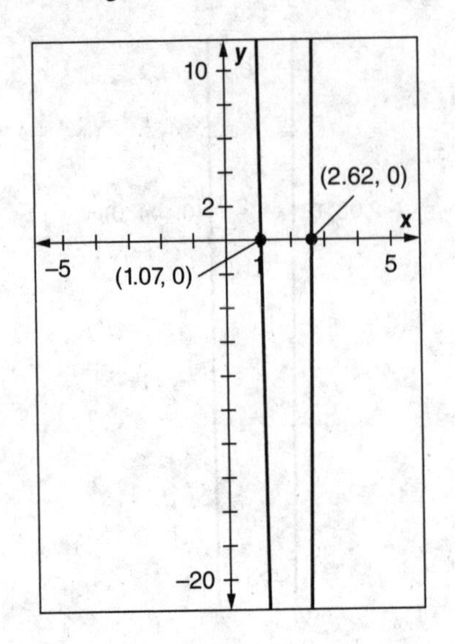

Converting the zeroes to fractions, you get $x = \frac{15}{14}$ and $x = \frac{21}{8}$. Changing these to the binomials $14x - 15 = 0$ and $8x - 21 = 0$, you can determine that $112x^2 - 414x + 315$ factors to $(8x - 21)(14x - 15)$.

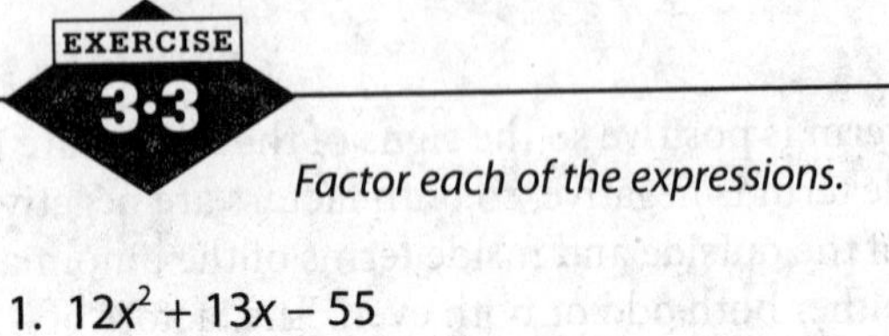

Factor each of the expressions.

1. $12x^2 + 13x - 55$
2. $48x^2 + 110x + 63$
3. $24x^2 + 22x - 65$
4. $36x^2 + 69x + 28$
5. $50x^2 - 105x + 54$
6. $54x^2 + 69x + 20$
7. $48x^2 + 8y - 225$
8. $24x^2 + 35x - 150$

Completing the square

The "square" in the process of completing the square is the square trinomial (which will have to be square if one chooses to provide a geometric model of the algebraic expression). Recall that $a^2 + 2ab + b^2 = (a + b)^2$ and $a^2 - 2ab + b^2 = (a - b)^2$. The middle terms are twice the square roots of the first and third terms. This is the key for completing the square: Take one half the coefficient of the linear term of the expression, square it, and add it.

PROBLEM What number must be added to $x^2 + 12x$ to complete the square?

SOLUTION Rewrite $x^2 + 12x$ as $x^2 + 2(6)x$ to recognize that 6 is the square root of the required third term to complete the square. Adding 36 gives $x^2 + 12x + 36$, which equals $(x + 6)^2$.

The problem asked to find the third number that would complete the square. Please be sure to recognize that the original expression $x^2 + 12x$ is not the same as the final answer $(x + 6)^2$ because 36 was added to the original expression.

PROBLEM Use the technique of completing the square to rewrite the equation of the parabola $f(x) = x^2 + 12x + 31$ in vertex form.

SOLUTION As seen in the previous example, 36 must be added to $x^2 + 12x$ to complete the square. Consequently, $f(x) = x^2 + 12x + 31$ can be written as $f(x) = x^2 + 12x + 36 - 36 + 31$, which can be simplified to $f(x) = (x + 6)^2 - 5$.

PROBLEM Rewrite $g(x) = x^2 - 5x - 3$ in vertex form.

SOLUTION Half of the linear coefficient is $\frac{5}{2}$ and the square of this number is $\frac{25}{4}$. Add and subtract $\frac{25}{4}$ to $g(x)$ to get $g(x) = \left(x^2 - 5x + \left(\frac{5}{2}\right)^2\right) - \frac{25}{4} - 3$. When simplified, $g(x) = \left(x - \frac{5}{2}\right)^2 - \frac{37}{4}$.

The process is a little more complicated if the quadratic coefficient is not 1. In that case, the first step is to factor the quadratic coefficient from the quadratic and linear terms and then proceed as usual.

PROBLEM Complete the square for $4x^2 - 24x - 9$.

SOLUTION Factor the quadratic coefficient to get $4(x^2 - 6x) - 9$. (Notice that the constant at the end of the problem remains unfactored and is not included within the parentheses.) Half of 6 is 3, so add 3^2 inside the parentheses to complete the square, and subtract $4(3^2)$ outside the parentheses to maintain equality: $4(x^2 - 6x + 3^2) - 4(3^2) - 9 = 4(x - 3)^2 - 45$.

PROBLEM Write the equation for the parabola $q(x) = \frac{-1}{3}x^2 - 4x + 3$ in vertex form.

SOLUTION Factor the quadratic coefficient to get $q(x) = \frac{-1}{3}(x^2 + 12x) + 3$. Half the linear coefficient is 6, so add 6^2 within the parentheses, and subtract $\frac{-1}{3}(6^2)$ outside the parentheses to balance the equation. $q(x) = \frac{-1}{3}(x^2 + 12x + 6^2) + \frac{1}{3}(6^2) + 3 = \frac{-1}{3}(x+6)^2 + 15$.

Completing the square can also be used to solve quadratic equations that cannot be factored.

PROBLEM Use completing the square to solve $4x^2 - 8x - 7 = 0$.

SOLUTION Because this is an equation, the balancing can be done by adding the same terms to both sides of the equation.

Add 7 to both sides: $4x^2 - 8x = 7$

Factor 4: $4(x^2 - 2x) = 7$

Complete the square: $4(x^2 - 2x + 1^2) = 7 + 4(1^2)$

Simplify: $4(x-1)^2 = 11$

Divide by 4: $(x-1)^2 = \frac{11}{4}$

Take the square root of both sides: $x - 1 = \pm\frac{\sqrt{11}}{2}$

Add 1: $x = 1 \pm \frac{\sqrt{11}}{2}$

EXERCISE
3·4

Rewrite the equations for each parabola in questions 1–4 in vertex form.

1. $f(x) = x^2 - 6x + 5$
2. $g(x) = 5x^2 + 15x - 3$
3. $p(x) = \frac{1}{4}x^2 + 5x - 6$
4. $q(x) = \frac{-2}{5}x^2 + 4x + 3$

Use completing the square to solve each of the equations in questions 5–8.

5. $x^2 + 8x - 5 = 0$
6. $4x^2 + 8x - 5 = 0$
7. $\frac{-3}{8}x^2 + 3x - 2 = 0$
8. $6x^2 + 13x + 6 = 0$

Quadratic formula

When applied to the general quadratic equation $ax^2 + bx + c = 0$, the process of completing the square yields the quadratic formula.

Move the constant to the right: $ax^2 + bx = -c$

Factor the quadratic coefficient: $a\left(x^2 + \frac{b}{a}x\right) = -c$

Complete the square: $a\left(x^2+\frac{b}{a}x+\left(\frac{b}{2a}\right)^2\right)=-c+a\left(\frac{b}{2a}\right)^2$

Simplify terms: $a\left(x+\frac{b}{2a}\right)^2=-c+\frac{b^2}{4a}=\frac{b^2-4ac}{4a}$

Divide by a: $\left(x+\frac{b}{2a}\right)^2=\frac{b^2-4ac}{4a^2}$

Take the square root: $x+\frac{b}{2a}=\pm\frac{\sqrt{b^2-4ac}}{2a}$

Solve for x: $x=\frac{-b\pm\sqrt{b^2-4ac}}{2a}$

PROBLEM Use the quadratic formula to solve $3x^2-5x-7=0$.

SOLUTION With $a=3$, $b=-5$, and $c=-7$, $x=\frac{-(-5)\pm\sqrt{(-5)^2-4(3)(-7)}}{2(3)}$. Please note the parentheses around the -5. A common error when using a calculator is to type $-5^2=-25$ rather than $(-5)^2=25$. $x=\frac{5\pm\sqrt{109}}{6}$.

PROBLEM Solve: $72x^2+25x-77=0$

SOLUTION This trinomial may or may not factor. The numbers are large enough that working through the options takes more time than it is worth—especially with the availability of the quadratic formula. Using $a=72$, $b=25$, and $c=-77$, $x=\frac{-25\pm\sqrt{(25)^2-4(72)(-77)}}{2(72)}=\frac{-25\pm\sqrt{22{,}801}}{144}=\frac{-25\pm151}{144}=\frac{126}{144},\frac{-176}{144}=\frac{7}{8},\frac{-11}{9}$.

Notice that this is the same result the zeroes of the graph gave when we looked at how to factor. The quadratic formula also can be used to help with factoring. $72x^2+25x-77$ factors to $(8x-7)(9x+11)$.

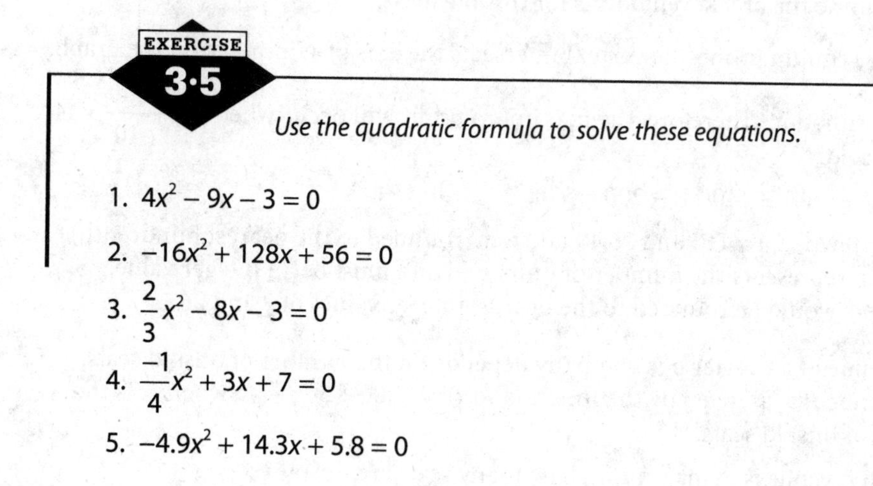

EXERCISE 3·5

Use the quadratic formula to solve these equations.

1. $4x^2-9x-3=0$
2. $-16x^2+128x+56=0$
3. $\frac{2}{3}x^2-8x-3=0$
4. $\frac{-1}{4}x^2+3x+7=0$
5. $-4.9x^2+14.3x+5.8=0$

Applications

When an object is thrown vertically into the air from the top of some height, the position of the object can be computed by the formula:

$$s(t) = \frac{-1}{2}gt^2 + v_0t + s_0$$

where g represents the acceleration constant due to gravity (32 ft/sec^2 in standard units; 9.8 m/sec^2 in metric units), v_0 is the initial velocity of the object, and s_0 is the initial height of the object.

PROBLEM A ball is thrown vertically from the top of a 50-ft-tall building with an initial upward velocity of 96 ft/sec.

a. Determine the maximum height of the ball.

b. Determine when the ball is at a height of 120 ft.

c. Determine when the ball hits the ground.

SOLUTION The data for the problem are given in standard units, so the equation for the position of the ball is $s(t) = -16t^2 + 96t + 50$.

a. The axis of symmetry for this parabola is $t = \frac{-96}{-32} = 3$. The ball is at a maximum height of 194 ft after 3 sec.

b. Substitute 120 for $s(t)$ to get the equation $120 = -16t^2 + 96t + 50$. Gather all terms to one side: $16t^2 - 96t + 70 = 0$. Use the quadratic formula to determine that $t = 0.85$ and 5.15 sec (answers rounded to the nearest hundredth). The ball reaches a height of 120 ft 0.85 sec after it is released while on the rise, and 5.15 sec after it is released on its descent.

c. The ball is at a height of 0 ft when it strikes the ground, so substitute 0 for $s(t)$ and solve the equation $0 = -16t^2 + 96t + 50$. Rounded to the nearest hundredth of a second, the solution to this problem is $t = 6.48$ sec. (The negative solution to this problem is ignored because the domain of this application is $t > 0$. Prior to this time, the ball has not been thrown.)

PROBLEM The weekly profit function for a small business is given by the equation $P(n) = -5n^2 + 140n - 200$, where n represents the number of units sold.

a. For what value of n is profit at the maximum?

b. What is the maximum profit?

c. The breakeven point for the business is the point at which the revenue taken in equals the expenditures, or the point at which the profit is equal to zero. Determine the breakeven points for this business.

SOLUTION a. The maximum profit (the vertex) intersects the axis of symmetry on the graph of the function. Therefore, the maximum profit will occur when $n = \frac{-140}{-10} = 14$ units sold.

b. $P(14) = -5(14)^2 + 140(14) - 200 = \$780$.

c. $P(n) = 0$ when $n = 1.51$ and 26.49 (answers rounded to the nearest hundredth). Since n represents the number of units sold and must be an integer value, answers would be rounded to the nearest integer values of 2 and 26.

PROBLEM The revenue of a charter bus company depends on the number of unsold seats. The revenue $R(x)$ is given by the function $R(x) = 5{,}000 + 50x - x^2$, where x is the number of unsold seats.

a. When revenue is at maximum, how many seats are unsold?

b. Find the maximum revenue.

c. What is the revenue when there are 10 unsold seats?

d. If the company wants a revenue of at least $5500, how many seats should it be willing to leave unsold?

SOLUTION a. The maximum revenue will occur at $x = \frac{-50}{-2} = 25$ seats left unsold.

b. $R(25) = \$5625$.

c. $R(10) = \$5400$.

d. The graph of the revenue function is concave down and the vertex is at the point (25, 5625). There is an interval between the two values of x for which the revenue exceeds $5500. Solve the equation $R(x) = 5{,}500$ to determine that $x = 13.8$ and 36.2 (answers rounded to the nearest tenth). Since x represents the number of seats left unsold and must be an integer value, the revenue will exceed $5,500 whenever the number of seats left unsold on the bus is between 14 and 36, inclusive.

EXERCISE
3·6

Solve each of the following problems.

A stone is thrown vertically into the air from the top of a cliff 40 ft above the ground with an initial upward velocity of 64 ft/sec.

1. What is the maximum height of the stone?
2. When is the stone at a height of 75 ft above the ground?
3. When does the stone hit the ground?

A ball is thrown vertically into the air from the roof of a building 15-m tall with an initial upward velocity of 19.6 m/sec.

4. What is the maximum height of the ball?
5. When is the ball at a height of 30 m?
6. When does the ball strike the ground?

A business determines that its revenue function for selling x items of its product is given by the equation $R(x) = -x^2 + 1200x$ and that the cost of producing these x items is given by the function $C(x) = 100x + 15{,}000$. Revenue and cost in dollars.

7. How many items should the company sell to maximize its revenue?
8. What is the maximum revenue?
9. How much does it cost to produce these items?
10. Profit is the difference between the revenue and cost. What is the profit for producing and selling these items?
11. Write an equation for the profit function.
12. At what point does the company begin to make a profit? When is the company working with a deficit (losing money)?
13. How many items must be produced and sold to maximize the profit?
14. What is the maximum profit?

Square root function

The graph of the equation $y = x^2$ passes the vertical line test ($y = x^2$ is a function) but fails the horizontal line test (the inverse of $y = x^2$ is not a function).

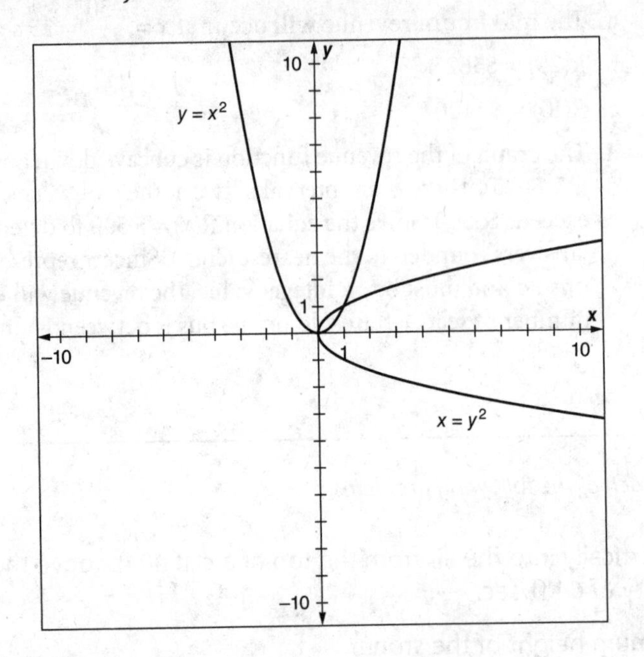

Yet, it is known that the square root function is the inverse of the function $y = x^2$. The explanation for this contradiction involves the important technique of restricting the domain. If the domain of the parabola $y = x^2$ is restricted to $x > 0$, the right-hand branch of the parabola passes both the vertical and horizontal line tests (making it a one-to-one function) and its inverse is the square root function.

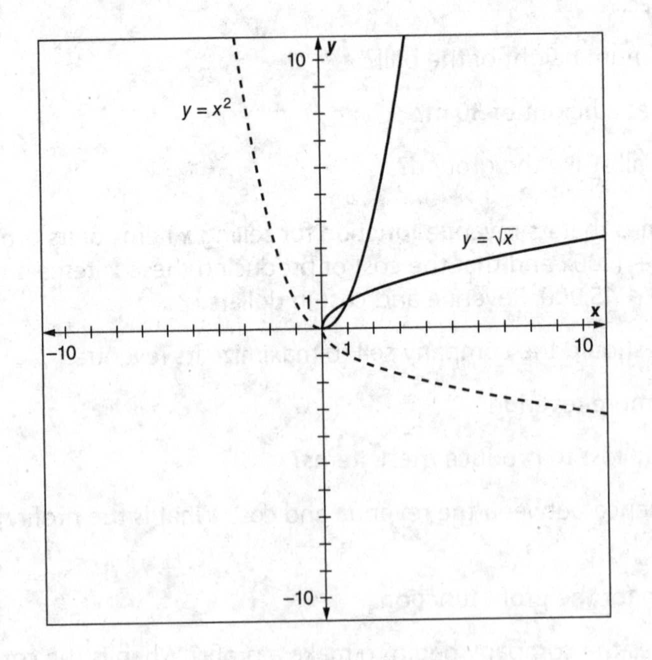

PROBLEM Given the parabola $y = 2(x+1)^2 - 3$, determine the restricted domain for which the function has an inverse and the equation of the inverse.

SOLUTION The vertex of the parabola is (–1, –3). Therefore, restricting the domain to $x \geq -1$ will yield the one-to-one function that has an inverse.

Interchange the x and y variables to find the inverse: $x = 2(y+1)^2 - 3$.

Add 3: $x + 3 = 2(y+1)^2$

Divide by 2: $\frac{1}{2}x + \frac{3}{2} = (y+1)^2$

Take the square root of both sides and subtract 1: $y = \sqrt{\frac{1}{2}x + \frac{3}{2}} - 1$.

This is the inverse of $y = 2(x+1)^2 - 3$. (Put each function into your graphing utility along with the graph of $y = x$ to see the symmetry that occurs with inverses. You can also examine a table of values to verify the outcome.)

EXERCISE 3·7

Determine the restricted domain of the given equations so that the function is one to one and provide the equation of the inverse function.

1. $y = \frac{1}{2}(x+3)^2 - 2$

2. $y = -3(x-5)^2 + 2$

Circles

By definition, a **circle** is the set of all points in a plane at a fixed distance r from a fixed point (h, k). Translating this definition into an equation, a point (x, y) is on the circle if its distance from (h, k) is equal to r, or $\sqrt{(x-h)^2 + (y-k)^2} = r$. Square both sides of the equation to get $(x-h)^2 + (y-k)^2 = r^2$. Of course, r is the length of the radius of the circle, and (h, k) are the coordinates for the center of the circle.

PROBLEM Write an equation for the circle centered at (–3, 6), which passes through the point (4, 5).

SOLUTION With the center being (–3, 6), the equation of the circle must take on the form $(x+3)^2 + (y-6)^2 = r^2$. Given that (4, 5) is a point on the circle, these values can be substituted to get $(4+3)^2 + (5-6)^2 = r^2$, or $r^2 = 50$. Therefore, the equation of the circle is $(x+3)^2 + (y-6)^2 = 50$.

PROBLEM When the binomials are expanded, the equation for a circle becomes $x^2 + y^2 - 8x - 12y - 29 = 0$. Find the coordinates of the center of the circle and the length of the radius of the circle.

SOLUTION Complete the square in both x and y to rewrite this equation (in general form) in center-radius form.

$$x^2 + y^2 - 8x - 12y - 29 = 0$$
$$x^2 - 8x + y^2 - 12y = 29$$

Add 16 to complete the square in x and 36 to complete the square in y.

$$x^2 - 8x + 16 + y^2 - 12y + 36 = 29 + 16 + 36$$
$$(x - 4)^2 + (y - 6)^2 = 81$$

The center of the circle is at (4, 6) and the length of the radius is 9.

EXERCISE

3·8

Solve the following problem.

1. Find the equation of the circle centered at (−5, 2), which passes through the point (4, 3).

Find the coordinates of the center and length of the radius for the circles whose equations are given in questions 2–5.

2. $(x + 4)^2 + (y - 9)^2 = 81$
3. $(x - 6)^2 + y^2 = 18$
4. $x^2 + y^2 + 14x - 20y - 20 = 0$
5. $x^2 + y^2 - 42x + 120y - 1000 = 0$

Ellipses

The algebraic definition of an **ellipse** is the set of all points in a plane with the property that the sum of the distances from two fixed points is a constant. The geometric definition is that an ellipse is the intersection of a plane and a cone in which the plane is at an angle between being parallel to an edge of the cone and parallel to the base of a cone. A good example of an ellipse is to hold a cup of water at an angle so that the water does not spill. The surface of the water in the cup will be in the shape of an ellipse.

The following figures show the graphs of an ellipse when the fixed points lie on the x-axis, when they lie on the y-axis, and the equations for each. In each case, the center of the ellipse is the origin.

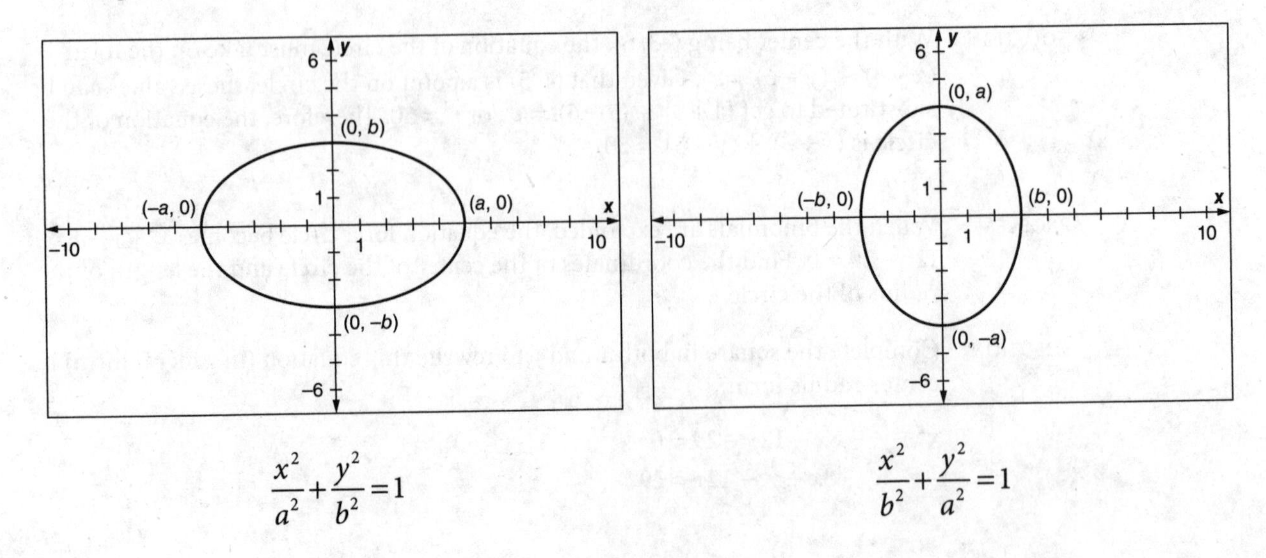

$$\frac{x^2}{a^2} + \frac{y^2}{b^2} = 1 \qquad\qquad \frac{x^2}{b^2} + \frac{y^2}{a^2} = 1$$

Observe that the longer axis (called the **major axis**) of the ellipse in each case has length $2a$, while the shorter (**minor**) axis has length $2b$. It is possible to determine the orientation of the major axis by examining the equation. The larger denominator represents the value of a^2, and the variable in the numerator reveals the orientation of the major axis.

If the center of the ellipse is translated from the origin to the point (h, k), then the equations of the graphs in standard form become

$$\frac{(x-h)^2}{a^2}+\frac{(y-k)^2}{b^2}=1 \text{ and } \frac{(x-h)^2}{b^2}+\frac{(y-k)^2}{a^2}=1, \text{ respectively.}$$

The ellipse is symmetric with respect to both the major and minor axes.

PROBLEM An ellipse with center (4, 8) is tangent to both the x-axis and the y-axis. Write an equation for the ellipse.

SOLUTION The center of the ellipse will be 4 units from the y-axis and 8 units from the x-axis. This makes the vertical axis the major axis with a length of 8, and the horizontal axis the minor axis with a length of 4. The equation of the ellipse is $\frac{(x-4)^2}{16}+\frac{(y-8)^2}{64}=1.$

PROBLEM Find the coordinates of the x-intercepts and y-intercepts of the ellipse with the equation $\frac{(x-3)^2}{16}+\frac{(y-4)^2}{9}=1.$

SOLUTION To find the x-intercepts, set $y=0$ and solve for x.

$\frac{(x-3)^2}{16}+\frac{(-4)^2}{9}=1$ becomes $\frac{(x-3)^2}{16}+\frac{16}{9}=1$ so that $\frac{(x-3)^2}{16}=\frac{-7}{9}$. Since there are no real numbers that, when squared, equal a negative number, this ellipse does not cross the x-axis.

To find the y-intercepts, set $x=0$ and solve for y.

$\frac{(-3)^2}{16}+\frac{(y-4)^2}{9}=1$ becomes $\frac{9}{16}+\frac{(y-4)^2}{9}=1$ so that $\frac{(y-4)^2}{9}=\frac{7}{16}$. Take the square root of both sides of the equation to get $\frac{y-4}{3}=\frac{\pm\sqrt{7}}{4}$. Multiply by 3 and add 4 to find that the y-intercepts are at $\left(0, 4+\frac{\pm 3\sqrt{7}}{4}\right)$.

PROBLEM The endpoints of the major axis of an ellipse are at (–3, 2) and (9, 2), and the ellipse passes through the point (0, –1). Determine the equation of the ellipse.

SOLUTION The segment containing the endpoints of the major axis is horizontal. The center of the ellipse is the midpoint of the major axis, (3, 2), and the length of the major axis is 12, making $a=6$. The equation of the ellipse must be of the form $\frac{(x-3)^2}{36}+\frac{(y-2)^2}{b^2}=1$. Because the ellipse passes through (0, –1), substitute 0 for x and –1 for y to solve for b^2. $\frac{(x+3)^2}{9}+\frac{(y+2)^2}{25}=1$ becomes $\frac{1}{4}+\frac{9}{b^2}=1$, or $b^2=12$. The equation of the ellipse is $\frac{(x-3)^2}{36}+\frac{(y-2)^2}{12}=1.$

PROBLEM Determine the coordinates of the center, the length of the major axis, and the length of the minor axis for the ellipse whose standard equation is:

$$25x^2 + 9y^2 - 150x + 72y + 144 = 0$$

SOLUTION Gather terms in x and y and move the constant to the right side of the equation:

$$25x^2 - 150x + 9y^2 + 72y = -144$$

Complete the square in x and y:

$$25(x^2 - 6x) + 9(y^2 + 8y) = -144$$

$$25(x^2 - 6x + 3^2) + 9(y^2 + 8y + 4^2) = -144 + 25(3^2) + 9(4^2)$$

$$25(x - 3)^2 + 9(y + 4)^2 = -144 + 225 + 144$$

$$25(x - 3)^2 + 9(y + 4)^2 = 225$$

Divide by 225: $$\frac{(x-3)^2}{9} + \frac{(y+4)^2}{25} = 1$$

The center of the ellipse is at $(3, -4)$, the major axis has a length of 10, and the minor axis has a length of 6.

EXERCISE

3·9

Solve the problems in this exercise as indicated.

1. The endpoints of the major axis of an ellipse have coordinates $(-2, 5)$ and $(-2, -7)$. If an endpoint of the minor axis has coordinates $(-5, -1)$, write an equation for the ellipse.
2. Find the coordinates of the x- and y-intercepts for an ellipse with the equation $\frac{(x+1)^2}{9} + \frac{(y-2)^2}{8} = 1$.
3. The endpoints of the major axis of an ellipse are located at $(2, -3)$ and $(12, -3)$. If the ellipse is tangent to the x-axis, write an equation for the ellipse.
4. Write, in standard form, the equation for the ellipse defined by $3x^2 + 5y^2 + 18x - 40y - 43 = 0$.

Hyperbolas

The algebraic definition of a **hyperbola** is the set of all points in a plane with the property that the absolute value of the differences of the distances from two fixed points is a constant. The geometric definition of a hyperbola is the intersection of a plane and both naps of a double-napped cone in which the plane is at an angle between being parallel to an edge of the cone and perpendicular to the bases of the double-napped cones. The outline of the cooling towers in nuclear reactors are in the shape of hyperbolas (the shape of the cooling tower is called a hyperboloid.)

Sketches of two hyperbolas, each centered at the origin, one with the vertices on the x-axis and the other with the vertices on the y-axis, along with their equations in standard form, are shown in the following images.

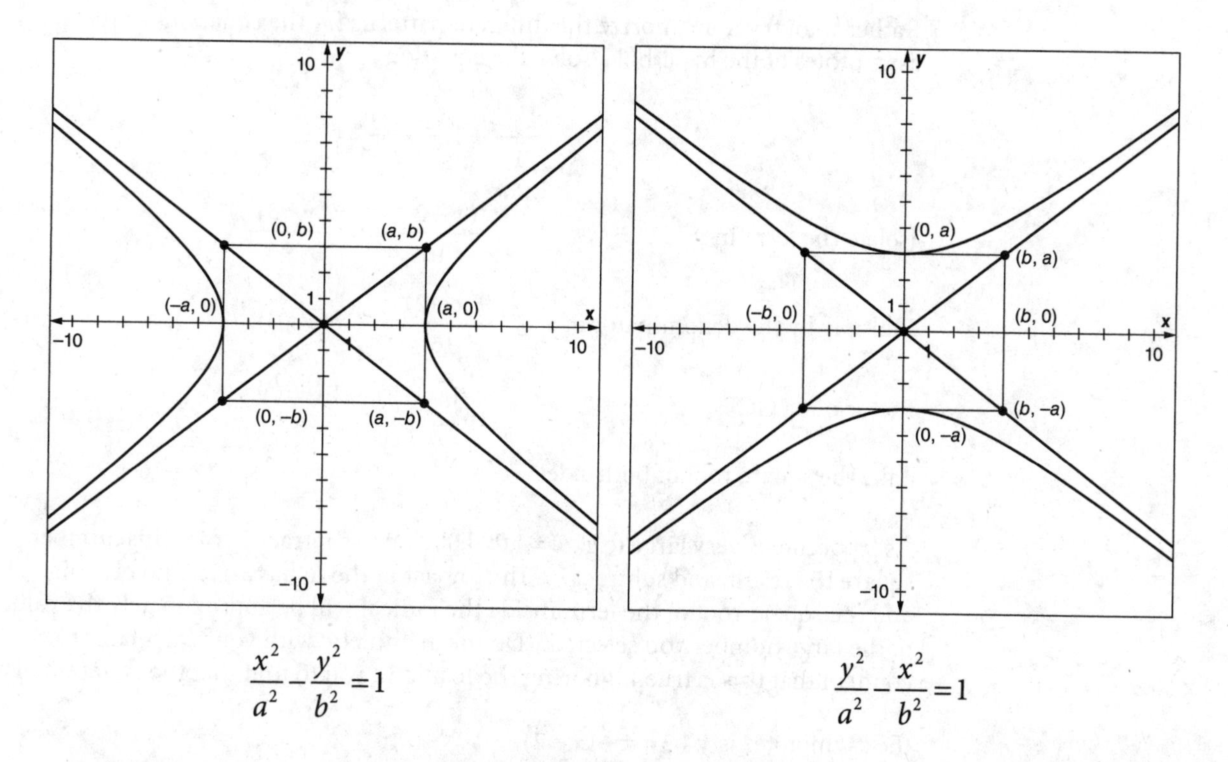

$$\frac{x^2}{a^2}-\frac{y^2}{b^2}=1 \qquad \frac{y^2}{a^2}-\frac{x^2}{b^2}=1$$

Observe that the variable in the numerator of the leading fraction indicates the direction in which the vertices of the hyperbola are located from the center. The first denominator is always a^2. The rectangle formed has dimensions $2a \times 2b$, and the diagonals of the rectangle are called the asymptotes. **Asymptotes** are lines that the hyperbola gets closer and closer to as x increases but never actually touches them. As the absolute value of x gets very large, the graph of the hyperbola becomes much closer to these lines. The asymptotes are very useful for graphing the hyperbolas, and also give an indication that if each of the equations for the hyperbola is solved to be in the form $y =$, the resulting equations will contain terms in x that become negligible as $|x|$ becomes very large.

If the center of the hyperbola is translated from the origin to (h, k), then the equations for the resulting hyperbolas will be:

$$\frac{(x-h)^2}{a^2}-\frac{(y-k)^2}{b^2}=1 \text{ and } \frac{(y-k)^2}{a^2}-\frac{(x-h)^2}{b^2}=1\text{, respectively.}$$

There is another equation for a special set of hyperbolas. Equations of the form $y=\frac{c}{x}$, where c is a constant, are hyperbolas that lie in Quadrants I and III when $c > 0$, and in Quadrants II and IV when $c < 0$.

PROBLEM Given the hyperbola with the equation $\frac{(x-2)^2}{16}-\frac{(y+3)^2}{9}=1$, find the coordinates of the center of the hyperbola, the coordinates of the vertices, and the equation of the asymptotes.

SOLUTION The center of the hyperbola is (2, –3). Since $a^2 = 16$ and the variable in the numerator is x, the vertices of this hyperbola will be $a = 4$ units to the left and right of center. The vertices will be at (–2, –3) and (6, –3).

Rather than try to memorize the different patterns for the equations of the asymptotes of the hyperbola, solve the equation for y:

$$\frac{(x-2)^2}{16}-\frac{(y+3)^2}{9}=1$$

Isolate the term in y: $$\frac{(x-2)^2}{16}-1=\frac{(y+3)^2}{9}$$

Multiply by the denominator of y: $$\frac{9(x-2)^2}{16}-9=(y+3)^2$$

Factor out a GCF: $$\frac{9}{16}((x-2)^2-16)=(y+3)^2$$

Take the square root of both sides: $$y+3=\pm\frac{3}{4}\sqrt{(x-2)^2-16}$$

As $|x|$ becomes very large (e.g., $x = 1{,}000{,}000{,}002$), subtract 2 from this number, square the result, and subtract 16. The impact of the 16 has *almost* no bearing, and the square root of the term inside the radical will be approximately the value of the large number you selected. (Do the arithmetic with your calculator to confirm that this is true.) Ignoring the 16 allows you to find that the equation of the asymptotes is $y+3=\pm\frac{3}{4}(x-2)$.

The net result of dropping the 16 in this problem allows you to simplify the process for finding the equations of the asymptotes—replace the 1 on the right-hand side of the equation with 0 and solve for y.

PROBLEM Given the hyperbola with equation $\frac{(y+3)^2}{9}-\frac{(x-2)^2}{16}=1$, find the coordinates of the center of the hyperbola, the coordinates of the vertices, and the equation of the asymptotes.

SOLUTION The center of this hyperbola is also (2, –3). Since the first denominator is $a^2 = 9$ and is under the term in y, the vertices of this hyperbola will be $a = 3$ units above and below the center. The vertices will have coordinates (2, –6) and (2, 0). Replace the 1 with a 0 on the right-hand side of the original equation and solve for y to determine the equation of the asymptotes. $\frac{(y+3)^2}{9}-\frac{(x-2)^2}{16}=0$ becomes $\frac{(y+3)^2}{9}=\frac{(x-2)^2}{16}$. This, in turn, becomes $(y+3)^2=\frac{9(x-2)^2}{16}$ or $y+3=\pm\frac{3}{4}(x-2)$.

PROBLEM Rewrite the equation $25x^2 - 9y^2 - 200x - 90y - 50 = 0$ in the standard form of the hyperbola.

SOLUTION This requires that the technique of completing the square must be applied to the general form of the hyperbola. When doing so, care must be taken because of the negative sign that will exist in front of one of the squared terms.

Gather terms: $25x^2 - 200x - 9y^2 - 90y = 50$

Factor the quadratic coefficients: $25(x^2 - 8x) - 9(y^2 + 10y) = 50$

Complete each square: $25(x^2 - 8x + 16) - 9(y^2 + 10y + 25) = 50 + 25(16) - 9(25)$

Simplify: $25(x-4)^2 - 9(y+5)^2 = 225$

Divide by 225: $\dfrac{(x-4)^2}{9} - \dfrac{(y+5)^2}{25} = 1$

EXERCISE 3·10

Solve the following problems.

1. Determine the coordinates of the center, the coordinates of the vertices, and the equations of the asymptotes of the hyperbola with equation $\dfrac{(x-4)^2}{6} - \dfrac{(y+5)^2}{24} = 1$.
2. Determine the coordinates of the center, the coordinates of the vertices, and the equations of the asymptotes of the hyperbola with equation $\dfrac{(y-4)^2}{6} - \dfrac{(x+5)^2}{24} = 1$.
3. Rewrite the equation $49y^2 - 16x^2 + 32x - 686y + 1601 = 0$ in the standard form of the hyperbola.
4. Rewrite the equation $81x^2 - 64y^2 + 324x + 1152y - 10{,}044 = 0$ in the standard form of the hyperbola.

Systems of equations

For the most part, the systems of equations you have been asked to solve up to this point have all involved linear equations. In this section, you will examine linear–quadratic and quadratic–quadratic systems of equations.

PROBLEM Solve the system of equations:

$$x^2 + y^2 = 25$$
$$y = x + 1$$

SOLUTION The linear equation is of the form $y =$. This is a good indication that substitution should be used to solve the system. Replace the y term in the equation of the circle with $x + 1$:

$$x^2 + (x+1)^2 = 25$$

Expand: $x^2 + x^2 + 2x + 1 = 25$

Gather like terms: $2x^2 + 2x - 24 = 0$

Factor: $2(x+4)(x-3) = 0$

Solve for x: $x = -4, 3$

Find y: $x = -4: y = -4 + 1 = -3$ $\quad x = 3: y = 3 + 1 = 4$

The points of intersection are $(-4, -3)$ and $(3, 4)$.

PROBLEM Solve the system of equations:

$$x^2 + y^2 = 25$$
$$y = x^2 - 5$$

SOLUTION Rewrite the equation of the parabola as $y + 5 = x^2$. A substitution for x^2 can be made in the equation of the circle to get $y + 5 + y^2 = 25$.

Gather terms: $y^2 + y - 20 = 0$

Factor: $(y + 5)(y - 4) = 0$

Solve for y: $y = -5, 4$

Solve for x: $y = -5$: $-5 + 5 = x^2$ yields $x = 0$ $\quad$ $y = 4$: $4 + 5 = x^2$ yields $x = \pm 3$

The points of intersection are $(0, -5)$, $(-3, 4)$, and $(3, 4)$.

PROBLEM Solve the system of equations:

$$\frac{x^2}{9} + \frac{y^2}{4} = 1$$
$$\frac{x^2}{4} + \frac{y^2}{8} = 1$$

SOLUTION Use a common denominator to remove the fractions from each equation:

$$4x^2 + 9y^2 = 36$$
$$2x^2 + y^2 = 8$$

Use the elimination method. Multiply the second equation by -9 and add the equations together.

$$4x^2 + 9y^2 = 36$$
$$-18x^2 - 9y^2 = -72$$

Add: $-14x^2 = -36$

Divide: $x^2 = \dfrac{36}{14}$

Solve for x: $x = \dfrac{\pm 6}{\sqrt{14}} = \dfrac{\pm 6\sqrt{14}}{14} = \dfrac{\pm 3\sqrt{14}}{7}$

Solve for y: With $x^2 = \dfrac{36}{14}$, $2\left(\dfrac{36}{14}\right) + y^2 = 8$ so $y^2 = \dfrac{20}{7}$ or $y = \pm\dfrac{2\sqrt{5}}{7}$

The points of intersection are $\left(\dfrac{3\sqrt{14}}{7}, \dfrac{2\sqrt{5}}{7}\right)$, $\left(\dfrac{3\sqrt{14}}{7}, \dfrac{-2\sqrt{5}}{7}\right)$, $\left(\dfrac{-3\sqrt{14}}{7}, \dfrac{-2\sqrt{5}}{7}\right)$,

and $\left(\dfrac{-3\sqrt{14}}{7}, \dfrac{2\sqrt{5}}{7}\right)$.

Clearly, the solutions to the systems of equations can become very "interesting" quickly. Don't let this scare you. Follow the process. Check your answers with your calculator, using decimal approximations if you need to feel more comfortable. Be warned that with the approximations, the checks might not be completely accurate because of round-off errors. Just know that if the result was supposed to be 8 and the calculator result is 8.000101, you did the problem correctly.

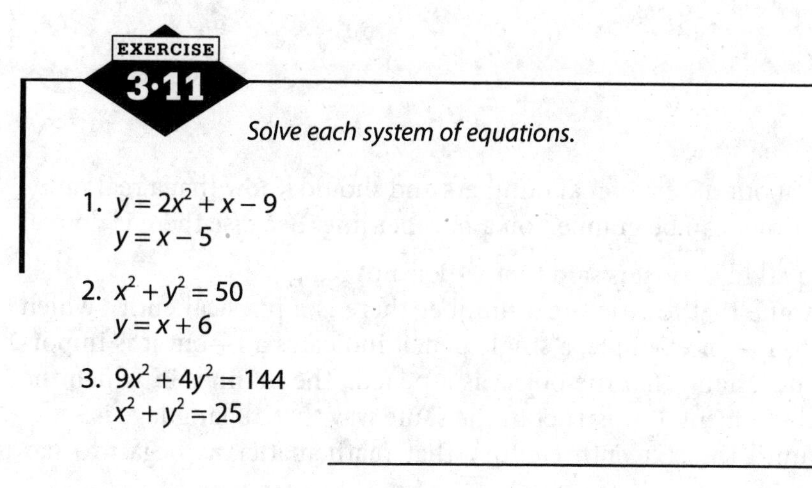

EXERCISE

3·11

Solve each system of equations.

1. $y = 2x^2 + x - 9$
   $y = x - 5$
2. $x^2 + y^2 = 50$
   $y = x + 6$
3. $9x^2 + 4y^2 = 144$
   $x^2 + y^2 = 25$

·4·

Complex numbers

You have learned about the set of real numbers and should know that a real number is any number that can be graphed on a number line. Because there is no real number that is equal to $\sqrt{-1}$, it is said that $\sqrt{-1}$ is not real.

Some will argue that for each real number there is a physical entity which can be matched to it—for example, a single pencil indicates a 1—but it is important for you to understand that the object is a pencil. The manner in which the item is quantified is a mental construct in the same way that all language is.

It was not until the sixteenth century that mathematicians began to pay greater attention to this number. Because $\sqrt{-1}$ was not real, the number was called **imaginary** and $\sqrt{-1} = i$.

Powers of *i*

If $\sqrt{-1} = i$, then $(\sqrt{-1})^2 = i^2$. The rules of arithmetic have always stated that $(\sqrt{a})^2 = a$ so $i^2 = -1$. Multiply both sides of this equation by i to get $i^3 = -1(i) = -i$. Do so again to find that $i^4 = 1$. A rather surprising and interesting consequence of this is that the powers of i will repeat themselves. That is, $i^5 = i^4 i = i$; $i^6 = i^4 i^2 = -1$; $i^7 = i^4 i^3 = -i$; $i^8 = i^4 i^4 = 1$. The rule is simple: when simplifying powers of i, divide the exponent by 4 and match the remainder with 0, 1, 2, or 3.

PROBLEM Simplify i^{156}.

SOLUTION $156 \div 4 = 39\ r0$. Therefore, $i^{156} = i^0 = 1$.

PROBLEM Simplify i^{239}.

SOLUTION $239 \div 4 = 59\ r3$. Therefore, $i^{239} = i^3 = -i$.

PROBLEM Simplify i^{-9}.

SOLUTION Negative exponents indicate reciprocals so $i^{-9} = \frac{1}{i^9}$. Multiply numerator and denominator by i^3 (which will make the denominator $i^{12} = 1$) to get $\frac{i^3}{i^{12}} = i^3 = -i$.

If $i^2 = -1$, then what is the value of $(4i)^2$? The rules of exponents tell you that $(4i)^2 = 16i^2 = 16(-1) = -16$.

If $\sqrt{-1} = i$, what is the value of $\sqrt{-25}$? The rules for simplifying square roots tell you that $\sqrt{-25} = \sqrt{-1}\sqrt{25} = 5i$.

The most difficult problem is $(\sqrt{-16})(\sqrt{-25})$. The temptation is to multiply −16 and −25 to get 400 and conclude that the square root of 400 is 20. However, the $\sqrt{-1}$ must first be factored from each of the terms. $(\sqrt{-16})(\sqrt{-25}) = (\sqrt{-1}\sqrt{16})(\sqrt{-1}\sqrt{25}) = (\sqrt{-1})^2(\sqrt{16}\sqrt{25}) = (-1)(4)(5) = -20$.

PROBLEM Simplify $(-7i)^2$.

SOLUTION $(-7i)^2 = 49i^2 = 49(-1) = -49$.

PROBLEM Simplify $(4i)^3$.

SOLUTION $(4i)^3 = 4^3i^3 = 64(-i) = -64i$.

PROBLEM Simplify $\sqrt{-18}$.

SOLUTION $\sqrt{-18} = \sqrt{-1}\sqrt{18} = \sqrt{-1}\sqrt{9}\sqrt{2} = 3i\sqrt{2}$. Notice that the i is written between the 3 and the radical so that it is not mistaken as part of the radicand.

PROBLEM Simplify $(\sqrt{-8})(\sqrt{-32})$.

SOLUTION $(\sqrt{-8})(\sqrt{-32}) = (\sqrt{-1}\sqrt{8})(\sqrt{-1}\sqrt{32}) = (\sqrt{-1})^2(\sqrt{4}\sqrt{2})(\sqrt{16}\sqrt{2})$
$= -(2\sqrt{2})(4\sqrt{2}) = -16$.

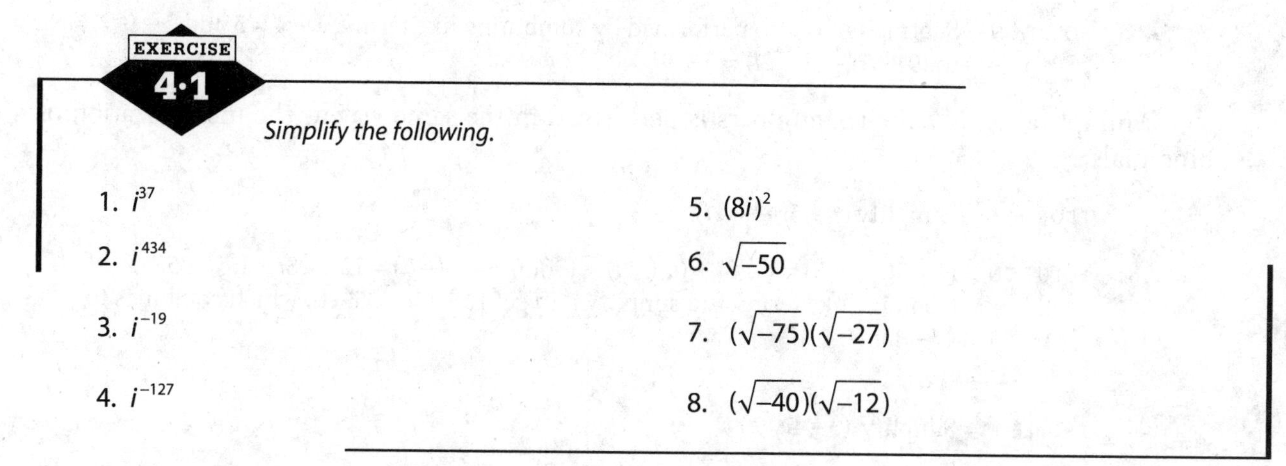

EXERCISE 4·1

Simplify the following.

1. i^{37}
2. i^{434}
3. i^{-19}
4. i^{-127}
5. $(8i)^2$
6. $\sqrt{-50}$
7. $(\sqrt{-75})(\sqrt{-27})$
8. $(\sqrt{-40})(\sqrt{-12})$

Arithmetic of complex numbers

If the imaginary numbers cannot fit on a number line, where do they exist geometrically? The answer is to extend the number line to a number plane. Through the origin of the real number line, construct a new number line perpendicular to it. The numbers of this line are the imaginary numbers. i, or $1i$, is one unit above the real number line while $-i$ is one unit below. The numbers that appear in the quadrants of this number plane are the complex numbers. A point with coordinates (4, 3) in the number plane equals $4 + 3i$. A number point with coordinates (–2, 5) equals $-2 + 5i$. The real numbers have an ordinate (second number in the ordered pair) of 0, while the imaginary numbers have an abscissa (first number in the ordered pair) of 0.

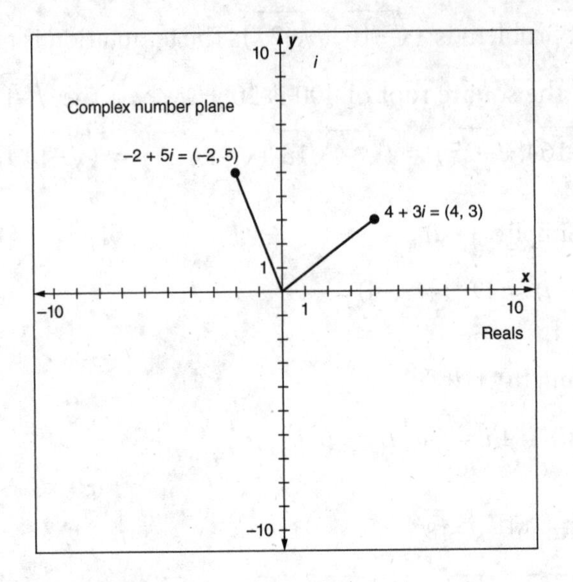

The sum of two complex numbers $a + bi$ and $c + di = (a + c) + (b + d)i$. When you combine the like parts, real numbers are added to real numbers and imaginary numbers are added to imaginary numbers.

PROBLEM Find the sum of $2 + 3i$ and $-5 + 8i$.

SOLUTION $2 + -5 = -3$ and $3 + 8 = 11$ so $(2 + 3i) + (-5 + 8i) = -3 + 11i$.

PROBLEM Subtract $4 - 2i$ from $9 + 7i$.

SOLUTION Subtraction is also performed by combining like terms. $9 - 4 = 5$ and $7 - (-2) = 9$ so $(9 + 7i) - (4 - 2i) = 5 + 9i$.

Multiplication of complex numbers is performed in the same way as the multiplication of binomials.

PROBLEM Simplify $(4 + 5i)(3 - 7i)$.

SOLUTION $(4 + 5i)(3 - 7i) = (4)(3) + (4)(-7i) + (5i)(3) + (5i)(-7i) = 12 - 28i + 15i - 35i^2$. Combine like terms and simplify i^2 to get $12 - 13i - 35(-1)$, which combines to $47 - 13i$.

PROBLEM Simplify $(3-4i\sqrt{2})^2$

SOLUTION The square of the complex number is computed in the same manner as the squared binomial. $(3-4i\sqrt{2})^2 = (3)^2-2(3)(4i\sqrt{2})+(4i\sqrt{2})^2 = 9 - 24i\sqrt{2} + 16i^2(2) = 9 - 24i\sqrt{2} - 32 = -23 - 24i\sqrt{2}$

The division of complex numbers requires the use of conjugates. Recall that $a+b$ and $a-b$ are called conjugates and that $(a+b)(a-b)=a^2-b^2$.

PROBLEM Simplify $\frac{10}{1+2i}$.

SOLUTION The conjugate of $1+2i$ is $1-2i$. Multiply the numerator and denominator of the fraction by $1-2i$ to get $\left(\frac{10}{1+2i}\right)\left(\frac{1-2i}{1-2i}\right)=\frac{10(1-2i)}{1^2-(2i)^2}$. Simplify the denominator to get $\frac{10(1-2i)}{1-4i^2}=\frac{10(1-2i)}{5}=2(1-2i)=2-4i$.

PROBLEM Simplify $\frac{4+3i\sqrt{2}}{2-5i\sqrt{2}}$.

SOLUTION Multiply the numerator and denominator of the fraction by $2+5i\sqrt{2}$, the conjugate of $2-5i\sqrt{2}$. $\left(\frac{4+3i\sqrt{2}}{2-5i\sqrt{2}}\right)\left(\frac{2+5i\sqrt{2}}{2+5i\sqrt{2}}\right)=\frac{(4+3i\sqrt{2})(2+5i\sqrt{2})}{2^2-(5i\sqrt{2})^2}=$ $\frac{8+20i\sqrt{2}+6i\sqrt{2}+15i^2(2)}{4-25i^2(2)}=\frac{8+26i\sqrt{2}-30}{4+50}=\frac{-22+26i\sqrt{2}}{54}=\frac{-11+13i\sqrt{2}}{27}$. Written in the true form of a complex number, $a+bi$, the answer is $\frac{-11}{27}+\frac{13\sqrt{2}}{27}i$.

Finally, two complex numbers, $a+bi$ and $c+di$, are equal if and only if $a=c$ and $b=d$.

PROBLEM Determine the values of a and b if $2a+1+3i=7-(2b-5)i$.

SOLUTION $2a+1=7$ yields $a=3$, and $3=-(2b-5)$ becomes $-3=2b-5$, so $b=1$.

EXERCISE 4·2

Perform each operation. Write your answers in a + bi *form.*

1. $(4+5i)+(9-3i)$
2. $(8+7i)-(-2+9i)$
3. $(2-3i)(4+5i)$
4. $(7+3i\sqrt{5})(4-2i\sqrt{5})$
5. $(7+3i\sqrt{5})^2$
6. $\frac{15}{3+4i}$
7. $\frac{5+2i\sqrt{3}}{3+i\sqrt{3}}$
8. Find the values of x and y if $(3x-2y)+(5x+7y)i=13+i$.

The discriminant and nature of the roots of a quadratic equation

With the inclusion of complex numbers, the set of solutions for equations is expanded. In the past, the solution to the equation $x^2=-1$ would have been "no real solution," and now the answer is $x=\pm i$.

PROBLEM Solve $x^2 + 6x + 25 = 0$.

SOLUTION Using the quadratic formula with $a = 1$, $b = 6$, and $c = 25$,

$$x = \frac{-6 \pm \sqrt{(6)^2 - 4(1)(25)}}{2(1)} = \frac{-6 \pm \sqrt{-64}}{2} = \frac{-6 \pm 8i}{2} = -3 \pm 4i.$$

PROBLEM Solve $4x^2 - 5x + 3 = 0$.

SOLUTION Using the quadratic formula with $a = 4$, $b = -5$, and $c = 3$,

$$x = \frac{-(-5) \pm \sqrt{(-5)^2 - 4(4)(3)}}{2(4)} = \frac{5 \pm \sqrt{-23}}{8} = \frac{5}{8} \pm \frac{\sqrt{23}}{8}i.$$

It is possible to tell the type of answers an equation will generate (also known as the **nature of the roots**) prior to working through the entire problem. The term $b^2 - 4ac$, the term within the square root when using the quadratic formula, is called the discriminant. The value of this number discriminates between real and complex solutions, equal and unequal solutions, and rational versus irrational solutions.

IF THE VALUE OF THE DISCRIMINANT IS	THE ROOTS ARE
0	real, rational, and equal
> 0 and a perfect square	real, rational, and unequal
> 0 and not a perfect square	real, irrational, and unequal
< 0	complex solutions

PROBLEM Determine the nature of the roots of the equation $5x^2 - 8x + 4 = 0$.

SOLUTION $b^2 - 4ac = (-8)^2 - (4)(5)(4) = 64 - 80 = -16$. The roots of the equation are complex numbers.

PROBLEM Determine the nature of the roots of the equation $25x^2 - 30x + 9 = 0$.

SOLUTION $b^2 - 4ac = (-30)^2 - (4)(25)(9) = 900 - 900 = 0$. The roots of the equation are real, rational, and equal. (These are also referred to as double roots.)

EXERCISE

4·3

For questions 1–4, compute the discriminant and use this result to determine the nature of the roots of the given equation.

1. $9x^2 - 24x + 16 = 0$
2. $5x^2 + 8x - 7 = 0$
3. $10x^2 + 9x + 4 = 0$
4. $24x^2 + 38x + 15 = 0$

For questions 5 and 6, solve the quadratic equations. Write your answers in a + bi form when appropriate.

5. $16x^2 - 16x + 13 = 0$
6. $64x^2 - 48x + 29 = 0$

Sum and product of roots of a quadratic equation

The solutions to the quadratic equation $ax^2 + bx + c = 0$ are $x = \frac{-b}{2a} + \frac{\sqrt{b^2 - 4ac}}{2a}$ and $x = \frac{-b}{2a} - \frac{\sqrt{b^2 - 4ac}}{2a}$. The sum of these roots is $\frac{-b}{a}$, while the product of these roots is $\frac{c}{a}$.

PROBLEM If one root of the quadratic equation $3x^2 + 6x + c = 0$ is -1, find the value of c.

SOLUTION One way to do this problem is to substitute -1 for x in the equation and solve for c. $3(-1)^2 + 6(-1) + c = 0$ becomes $3 - 6 + c = 0$, so that $c = 3$.

A second way to do this problem is to observe that the sum of the roots is $\frac{-6}{3} = -2$. With one root equal to -1, the other root must be -1. The product of the roots $(-1)(-1) = \frac{c}{3}$, so $c = 3$.

For this problem, it appears to be easier to substitute the root for x and find the value of c. Is this true for the next problem also?

PROBLEM If one root of the quadratic equation $3x^2 + 6x + c = 0$ is $\frac{-5}{2}$, what is the value of c?

SOLUTION Substitute $\frac{-5}{2}$ for x: $3\left(\frac{-5}{2}\right)^2 + 6\left(\frac{-5}{2}\right) + c = 0$, becomes $\frac{75}{4} - 15 + c = 0$ so that $c = \frac{-15}{4}$.

Alternatively, the sum of the roots is -2. $\frac{-5}{2} + r = -2$, so $r = \frac{1}{2}$. The product of the roots $\left(\frac{-5}{2}\right)\left(\frac{1}{2}\right) = \frac{c}{3}$, so that $c = \frac{-15}{4}$. For some students, this might be the easier approach.

PROBLEM If 9 is a root of the equation $4x^2 + 8x + c = 0$, find the other root.

SOLUTION The sum of the roots is $\frac{-8}{4} = -2$. Since 9 is one root, the other root must be -11.

Notice that the two roots of the quadratic equation, $x = \frac{-b}{2a} + \frac{\sqrt{b^2 - 4ac}}{2a}$ and $x = \frac{-b}{2a} - \frac{\sqrt{b^2 - 4ac}}{2a}$, are conjugates of each other. If the coefficients of the quadratic are integers, then irrational roots and complex roots will always occur in conjugate pairs. This makes writing a quadratic equation for a given set of roots much easier to do.

PROBLEM Write a quadratic equation with integral coefficients whose roots are $\frac{2}{3} + \frac{4\sqrt{5}}{3}$ and $\frac{2}{3} - \frac{4\sqrt{5}}{3}$.

SOLUTION The sum of the roots is $\frac{4}{3} = \frac{-b}{a}$ and the product of the roots is $\left(\frac{2}{3} + \frac{4\sqrt{5}}{3}\right)\left(\frac{2}{3} - \frac{4\sqrt{5}}{3}\right) = \frac{4}{9} - \frac{80}{9} = \frac{-76}{9} = \frac{c}{a}$. Because the denominators of the two results are different, the sum of the roots must be rewritten as an equivalent fraction with a denominator of 9: $\frac{-b}{a}=\frac{12}{9}$. Using $a=9$, $b=-12$, and $c=-76$, the equation becomes $9x^2-12x-76=0$. (Remember: you were directed to write an *equation*; $9x^2-12x-76$ is not an equation but an expression. Don't forget to include $=0$ as part of your answer.)

PROBLEM Write a quadratic equation with integral coefficients whose roots are $\frac{5}{4}+\frac{3\sqrt{5}}{2}i$ and $\frac{5}{4}-\frac{3\sqrt{5}}{2}i$.

SOLUTION The sum of the roots is $\frac{5}{2}=\frac{-b}{a}$, while the product of the roots is $\left(\frac{5}{4}-\frac{3\sqrt{5}}{2}i\right)\left(\frac{5}{4}+\frac{3\sqrt{5}}{2}i\right)=\left(\frac{5}{4}\right)^2-\left(\frac{3\sqrt{5}}{2}i\right)^2=\frac{25}{16}+\frac{45}{4}=\frac{205}{16}=\frac{c}{a}$. Rewrite the sum as $\frac{40}{16}$. Therefore, $a=16$, $b=-40$, and $c=205$. The equation is $16x^2-40x+205=0$.

EXERCISE 4·4

Solve questions 1–3 as indicated.

1. Determine the sum and product of the roots of the quadratic equation $2x^2+7x-3=0$.
2. If -5 is a root of the equation $6x^2+12x+c=0$, what is the other root?
3. Determine the value of c in the equation $4x^2-9x+c=0$ if one root of the equation is $\frac{5}{4}$.

For questions 4–6, write a quadratic equation with integral coefficients whose roots are given.

4. $\frac{5}{4}, \frac{-7}{9}$
5. $\frac{5}{3}+\frac{-7\sqrt{2}}{9}, \frac{5}{3}-\frac{-7\sqrt{2}}{9}$
6. $\frac{5}{8}-\frac{3\sqrt{2}}{4}i, \frac{5}{8}+\frac{3\sqrt{2}}{4}i$

Polynomial functions

The study of polynomial functions is an important part of the study of all functions. A **polynomial function** is a function of the form

$$f(x) = a_n x^n + a_{n-1}x^{n-1} + \cdots + a_1 x + a_0,$$

Where a_n, a_{n-1}, …, a_1, a_0 are real numbers and n is a nonnegative integer. The domain of $f(x)$ consists of all real numbers. Learning about roots, zeroes, and end behavior will be beneficial when looking at future topics.

Even and odd functions

The graphs of $f(x) = x^n$ when n is an even integer have two properties: they are symmetric to the y-axis, and they have the range $y \geq 0$. As n gets larger, the graph seems to flatten between –1 and 1 and grow more rapidly when $x < -1$ or $x > 1$. As the magnitude of x becomes extremely large (that is, as x goes to infinity or negative infinity), the graph of $f(x) = x^n$ goes to infinity.

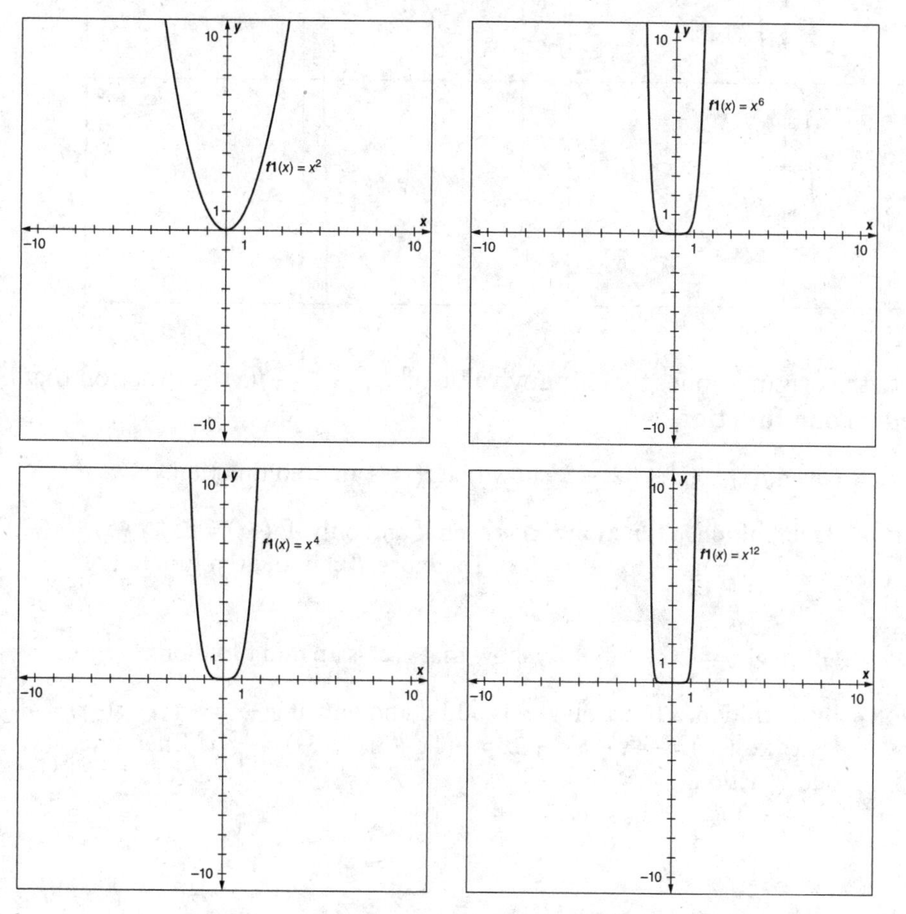

Symmetry to the y-axis implies that for any value of x, $f(-x) = f(x)$. A function that has this property is called an **even function.**

The graphs of $f(x) = x^n$ when n is an odd integer have two properties: they are symmetric to the origin and they have the set of real numbers as the range. As n gets larger, the graph seems to flatten between -1 and 1 and grow more rapidly when $x < -1$ or $x > 1$. As x goes toward negative infinity, the graph goes to negative infinity, and as x goes toward infinity, the graph goes to infinity.

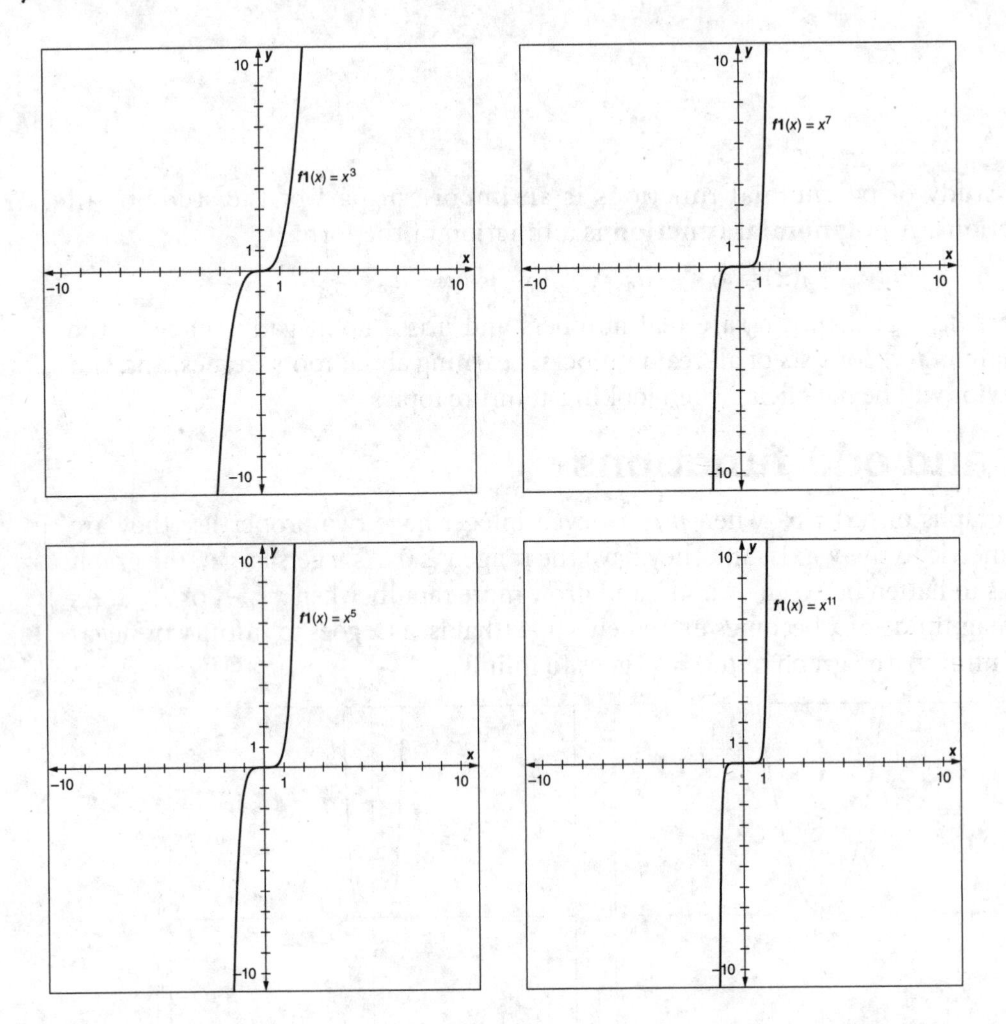

Symmetry to the origin implies that for any value of x, $f(-x) = -f(x)$. A function that has this property is called an **odd function.**

PROBLEM Given $f(x) = 4x^6 + 8x^4 + 3$, show that $f(x)$ is an even function.

SOLUTION By definition, a function $f(x)$ is even if and only if $f(-x) = f(x)$. $f(-x) = 4(-x)^6 + 8(-x)^4 + 3 = 4x^6 + 8x^4 + 3 = f(x)$. Therefore, $f(x)$ is an even function.

PROBLEM Given $g(x) = 4x^9 + 8x^3 + 3x$, show that $g(x)$ is an odd function.

SOLUTION By definition, a function $g(x)$ is odd if and only if $g(-x) = -g(x)$. $g(-x) = 4(-x)^9 + 8(-x)^3 + 3(-x) = -4x^9 - 8x^3 - 3x = -(4x^9 + 8x^3 + 3x) = -g(x)$. Therefore, $g(x)$ is an odd function.

PROBLEM Given $k(x) = 4x^9 + 8x^3 + 3$, is $k(x)$ an even function, an odd function, or neither?

SOLUTION Check $k(-x)$ to answer this question. $k(-x) = 4(-x)^9 + 8(-x)^3 + 3 = -4x^9 - 8x^3 + 3$. This is neither $k(x)$ nor $-k(x)$, so $k(x)$ is neither even nor odd.

An easy way to examine even versus odd functions is to examine the exponents of the terms that make up the polynomial. If all the exponents are odd, the function is an odd function, and if all the exponents are even, the function is even. (Recall that $x^0 = 1$ so that any constant, c, that appears at the end of a polynomial can be thought of as cx^0.)

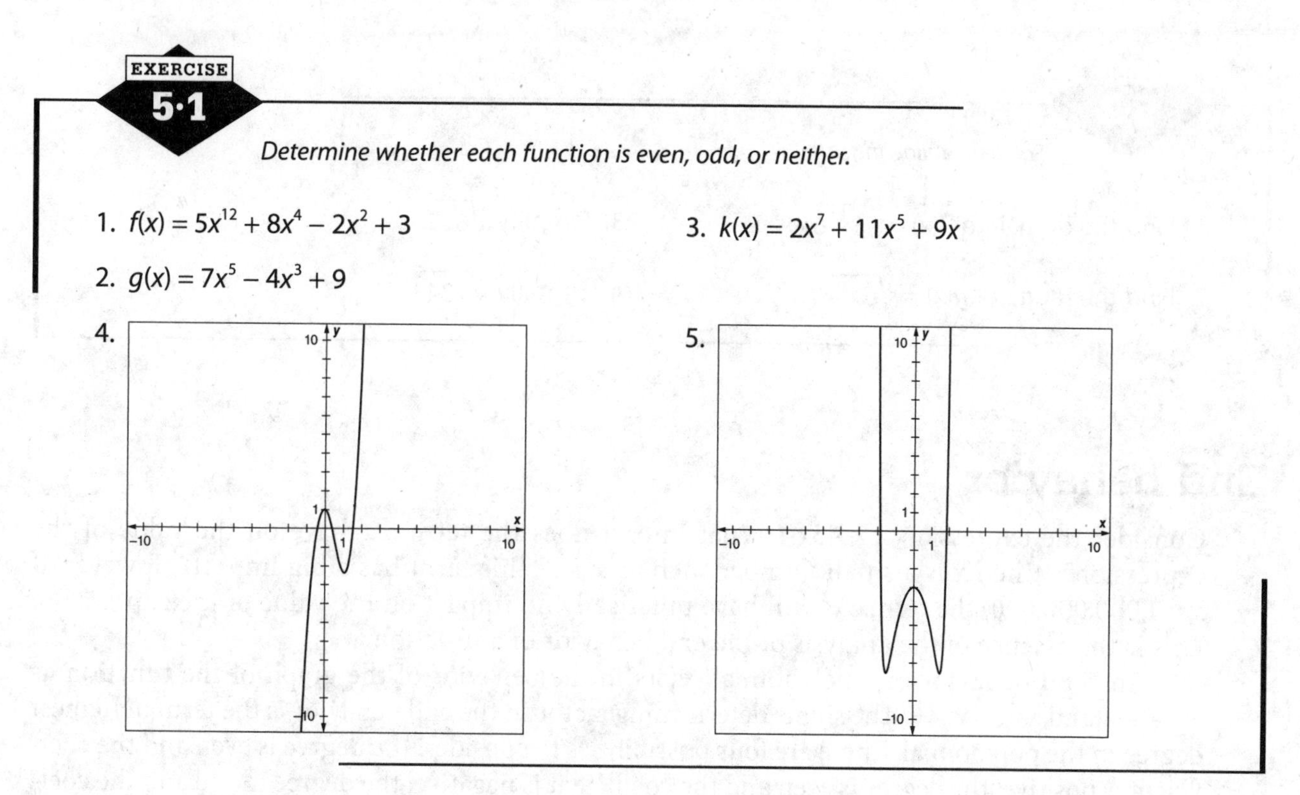

EXERCISE 5·1

Determine whether each function is even, odd, or neither.

1. $f(x) = 5x^{12} + 8x^4 - 2x^2 + 3$
2. $g(x) = 7x^5 - 4x^3 + 9$
3. $k(x) = 2x^7 + 11x^5 + 9x$
4.
5.

Inverse functions

Given $f(x) = x^n$, if n is even, the graph fails the horizontal line test so $f(x)$ would not have an inverse that is a function. If n is odd, the graph passes the horizontal line test and would have an inverse that is a function. The inverse of $f(x) = x^n$ is $f^{-1}(x) = \sqrt[n]{x}$. The domain of $f^{-1}(x) = \sqrt[n]{x}$ is $x \geq 0$ when n is even, and all real numbers when n is odd. The range of $f^{-1}(x) = \sqrt[n]{x}$ is $y \geq 0$ when n is even, and the set of real numbers when n is odd. While n is called the exponent in $f(x) = x^n$, it is referred to as the index for $f^{-1}(x) = \sqrt[n]{x}$.

PROBLEM Find the domain of the function $g(x) = \sqrt[4]{3x - 2}$.

SOLUTION Because the index is 4, the radicand (the expression inside the radical) must be nonnegative. Solve $3x - 2 \geq 0$ to get the domain $x \geq \frac{2}{3}$.

PROBLEM Simplify $\sqrt[3]{-64}$.

SOLUTION Because the index is 3, negative numbers are allowed inside the radical. The number, which when cubed is equal to −64, is −4.

PROBLEM Simplify $\sqrt[5]{-1024}$.

SOLUTION The solution of $x^5 = -1024$ is $x = -4$.

EXERCISE 5·2

Solve the following.

1. Find the domain of $f(x) = \sqrt{3-6x}$.
2. Find the range of $g(x) = \sqrt[3]{6x-3}$.
3. Simplify $\sqrt[4]{625}$.
4. Simplify $\sqrt[3]{-343}$.

End behavior

Consider the expression $x^4 - 5x^2$. What impact does the term $5x^2$ have on the value of the expression? When x is a small number such as $x = 2$, this term has a big impact. However, if $x = 1{,}000{,}000{,}000$, the term $5x^2$ will have practically no impact on the value of the expression. This is the essence of the analysis of the end behavior of a function.

The **end behavior** of a polynomial refers to the behavior of the graph of the function as $x \to -\infty$ and as $x \to +\infty$. The single determining factor in the end behavior is the term of highest degree in the polynomial. There are four possibilities to consider: the degree is even and the coefficient is positive; the degree is even and the coefficient is negative; the degree is odd and the coefficient is positive; the degree is odd and the coefficient is negative.

DEGREE	COEFFICIENT	$x \to -\infty$	$x \to +\infty$
Even	Positive	∞	∞
Even	Negative	$-\infty$	$-\infty$
Odd	Positive	$-\infty$	∞
Odd	Negative	∞	$-\infty$

PROBLEM Describe the end behavior of the function $f(x) = 8x^5 - 1200x^2 + 80$.

SOLUTION The degree of the polynomial is odd (5), and the coefficient of the fifth degree term is positive. Therefore, as $x \to -\infty$, $f(x)$ will go to $-\infty$ and as $x \to +\infty$, $f(x)$ will go to ∞.

PROBLEM Describe the end behavior of the function $g(x) = -4x^6 + 81x^5 + 19x^2 + 190$.

SOLUTION The degree of the polynomial is even (6), and the coefficient of the sixth degree term is negative. Therefore, as $x \to -\infty$, $g(x)$ will go to $-\infty$ and as $x \to +\infty$, $g(x)$ will go to $-\infty$. Use a graphing utility to verify the description of the end behaviors for these functions.

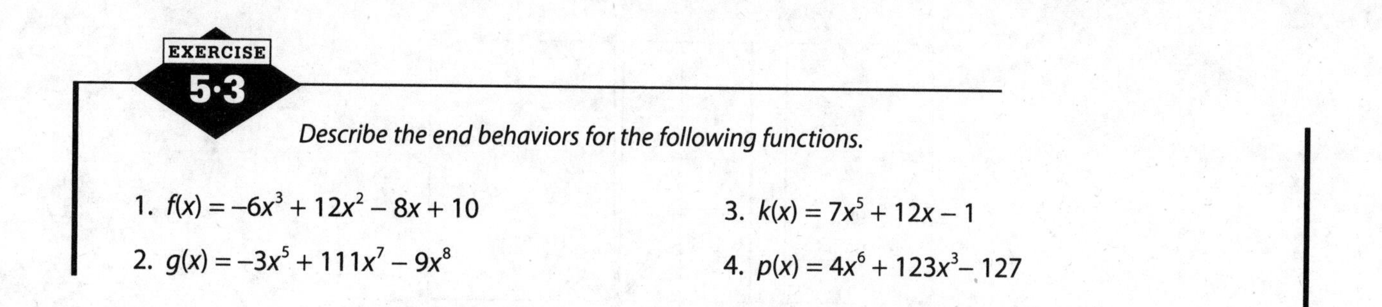

EXERCISE

5·3

Describe the end behaviors for the following functions.

1. $f(x) = -6x^3 + 12x^2 - 8x + 10$
2. $g(x) = -3x^5 + 111x^7 - 9x^8$
3. $k(x) = 7x^5 + 12x - 1$
4. $p(x) = 4x^6 + 123x^3 - 127$

Factor theorem

When one number is divided by a second number, the result is a quotient plus a remainder, understanding that it is possible that the remainder could be 0. In fact, when the remainder is 0, it is said that the second number (the divisor) is a factor of the first number (the dividend).

The factor theorem is an extension of this concept. Suppose a polynomial function $P(x)$ is divided by a linear term of the form $x - c$. The result will be a quotient $Q(x)$ and a remainder r. (Since the divisor is a linear term, the remainder must be a constant—if it was of a higher degree, it could be divided by the divisor.) As an equation, this is written as $\frac{P(x)}{x-c} = Q(x) + \frac{r}{x-c}$. Multiply both sides of this equation by the denominator $x - c$ to get the equation $P(x) = Q(x)\,(x - c) + r$.

Evaluate the function $P(x)$ with $x = c$: $P(c) = Q(c)\,(c - c) + r$. The value of the function when $x = c$ is the remainder when $P(x)$ is divided by $x - c$ (this is called the **remainder theorem**). Of more importance, if $P(x) = 0$, then $x - c$ is a factor of $P(x)$. This last sentence is the **factor theorem**.

PROBLEM Determine if $x - 3$ is a factor of $2x^3 + 7x^2 - 38x - 3$.

SOLUTION Using $P(x) = 2x^3 + 7x^2 - 38x - 3$ and $c = 3$, $P(3) = 2(3)^3 + 7(3)^2 - 38(3) - 3 = 0$. Yes, $x - 3$ is a factor of $2x^3 + 7x^2 - 38x - 3$.

PROBLEM Determine if $x + 4$ is a factor of $-2x^4 + 7x^3 + 8x^2 + 410$.

SOLUTION Using $P(x) = -2x^4 + 7x^3 + 8x^2 + 410$ and $c = -4$, $P(-4) = -2(-4)^4 + 7(-4)^3 + 8(-4)^2 + 410 = -422$. $x + 4$ is not a factor of $-2x^4 + 7x^3 + 8x^2 + 410$.

PROBLEM Determine if $2x + 3$ is a factor of $6x^3 - 13x^2 - 73x - 60$.

SOLUTION The factor theorem states that the linear factor should be of the form $x - c$, not $ax - c$. Rather than substituting c into $P(x)$, it is necessary to substitute $\frac{c}{a}$ (the solution to $ax - c = 0$) for x. With $P(x) = 6x^3 - 13x^2 - 73x - 60$, $P(-\frac{3}{2}) = 0$. Yes, $2x + 3$ is a factor of $6x^3 - 13x^2 - 73x - 60$.

PROBLEM Factor $120x^4 + 122x^3 - 601x^2 - 267x + 756$.

SOLUTION Use the graph of $P(x) = 120x^4 + 122x^3 - 601x^2 - 267x + 756$ to find the zeroes of the function. Once known, each of these values can be converted into a binomial factor of $120x^4 + 122x^3 - 601x^2 - 267x + 756$.

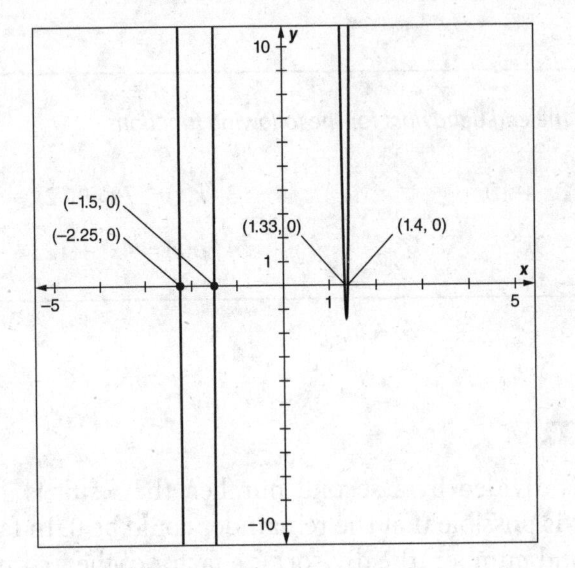

When converted to fractions, the zeroes of the function are $x = \frac{-9}{4}, \frac{-3}{2}, \frac{4}{3}, \frac{7}{5}$.

Each of these numbers came from the zero product property and the equations $4x + 9 = 0$, $2x + 3 = 0$, $3x - 4 = 0$, and $5x - 7 = 0$. Multiplied together, this becomes $(4x + 9)(2x + 3)(3x - 4)(5x - 7) = 0$. Consequently, $120x^4 + 122x^3 - 601x^2 - 267x + 756 = (4x + 9)(2x + 3)(3x - 4)(5x - 7)$.

EXERCISE 5·4

Solve the following.

1. Determine if $x - 5$ is a factor of $6x^3 - 13x^2 - 73x - 60$.
2. Determine if $4x + 3$ is a factor of $60x^3 + 137x^2 - 43x - 84$.
3. Factor $60x^3 + 137x^2 - 43x - 84$.
4. Factor $240x^4 + 214x^3 - 2053x^2 + 843x + 1260$.

Rational and irrational functions

The total resistance of resistors placed in series in an electrical circuit and the length of the period of a pendulum are examples of rational and irrational functions. In this chapter you will review the properties of these types of functions, the arithmetic processes for rational and irrational expressions, and the techniques for solving rational and irrational equations.

Rational functions

If $f(x)$ is equal to the ratio of two polynomial functions, $g(x)$ and $k(x)$ (i.e., $f(x)=\frac{g(x)}{k(x)}$), a number of questions arise. For what values of x is $f(x)$ undefined in the domain of real numbers? For what values of x does $f(x) = 0$ (the zeroes)? What is the behavior of $f(x)$ when the magnitude of the values of x become large (end behavior)? And what are the possible outcomes when $f(x)$ is evaluated (the range)?

For what values of x is $f(x)$ undefined? Recall that any nonzero number divided by zero is undefined. Any value of x that makes the denominator equal to zero must be rejected.

PROBLEM Determine the domain restrictions of $f(x)=\frac{3x+2}{x^2-5x-6}$.

SOLUTION Set $x^2 - 5x - 6 = 0$, factor to get $(x - 6)(x + 1) = 0$, and solve to get $x = 6, -1$. Therefore the domain restrictions of $f(x)$ are $x \neq 6, -1$.

PROBLEM Determine the domain restrictions of $v(x)=\frac{x}{x^2-3x+2}$.

SOLUTION Set $x^2 - 3x + 2 = 0$, factor to get $(x - 2)(x - 1) = 0$, and solve to get $x = 2, 1$. Therefore the domain restrictions of $v(x)$ are $x \neq 2, 1$.

For what values of x will $f(x) = 0$? A rational expression is equal to 0 when the numerator is 0 and the denominator is not.

PROBLEM Determine the zeroes of $f(x)$ if $f(x)=\frac{3x+2}{x^2-5x+6}$.

SOLUTION Solve $3x + 2 = 0$ to get $x=\frac{-2}{3}$.

PROBLEM Determine the zeroes of $g(x)$ if $g(x)=\frac{x^2-4}{x^2-3x+2}$.

SOLUTION Solve $x^2 - 4 = 0$ to get $x = \pm 2$. There is a trick to this problem, however. $x = 2$ is not an element in the domain of $g(x)$, so it cannot be included as a zero of the function. $g(x) = 0$ when $x = -2$.

There are a number of different possibilities for the end behavior of a rational function. For the level of the Algebra II curriculum, you will need to consider three cases: (1) the degree of the numerator is less than the degree of the denominator; (2) the degree of the numerator equals the degree of the denominator; and (3) the degree of the numerator is greater than the degree of the denominator. In each case, a simple explanation is to consider substituting a very large value for x in the function. Only the terms of largest degree in the numerator and denominator will matter, and the rest of the terms in the function can be ignored.

PROBLEM Determine the end behavior of $f(x)=\dfrac{3x+2}{x^2-5x-6}$.

SOLUTION Ignoring all but the terms of highest degree in the numerator and denominator, $f(x)$ is approximated by $\dfrac{3x}{x^2}$ which reduces to $\dfrac{3}{x}$. As x gets infinitely large, this expression gets closer to zero. The end behavior for $f(x)$ is that as $|x| \to \infty, f(x) \to 0$.

PROBLEM Determine the end behavior for $g(x)=\dfrac{x^2-4}{x^2-3x+2}$.

SOLUTION Ignoring all but the highest terms of the numerator and denominator, $g(x)$ is approximated by $\dfrac{x^2}{x^2}$ which reduces to 1. The end behavior for $g(x)$ is that as $|x| \to \infty, g(x) \to 1$.

PROBLEM Determine the end behavior for $q(x)=\dfrac{5x+3}{2x-7}$.

SOLUTION Continuing with the same method used in the previous two examples, $q(x)$ can be approximated by $\dfrac{5x}{2x}$ which reduces to $\dfrac{5}{2}$. The end behavior for $q(x)$ is that as $|x| \to \infty, q(x) \to \dfrac{5}{2}$.

PROBLEM Determine the end behavior for $v(x)=\dfrac{x^2+9}{5x+2}$.

SOLUTION This function can be approximated by $\dfrac{x^2}{5x}$ which reduces to $\dfrac{x}{5}$. The end behavior for $v(x)$ is that as $x \to -\infty$, $v(x) \to -\infty$, and as $x \to \infty$, $v(x) \to \infty$.

In general, if $f(x)=\dfrac{ax^m+\ldots}{bx^n+\ldots}$, the end behavior for $f(x)$ is: (1) if $m < n$, the graph goes to 0; (2) if $m = n$, the graph goes to $y=\dfrac{a}{b}$; and (3) if $m > n$, the graph gets infinitely large.

What are the possible outcomes when $f(x)$ is evaluated? This is a difficult question to answer because of the variety of rational functions you might encounter. You have seen from your study of inverse functions that the domain of the inverse function is the range of the original function. Computing the inverse of a rational function whose numerator and denominator are both linear expressions can be done. Other situations are not as straightforward, and the use of graphing technology is helpful for determining these ranges.

PROBLEM Find the inverse of the function $q(x)=\dfrac{5x+3}{2x-7}$.

SOLUTION If $q(x)$ is replaced with y, then $q(x)$ becomes $y=\dfrac{5x+3}{2x-7}$, and $q^{-1}(x)$ becomes $x=\dfrac{5y+3}{2y-7}$. Multiply by $2y-7$ to get $x(2y-7)=5y+3$. It is important for you to remember that you are trying to solve for y at this point. This will help you focus on the steps that you will need to take.

Distribute the x through the left side of the equation:	$2xy - 7x = 5y + 3$
Gather the terms in y on the left side:	$2xy - 5y = 7x + 3$
Factor out the y:	$y(2x - 5) = 7x + 3$
Divide by $2x - 5$:	$y = q^{-1}(x) = \dfrac{7x+3}{2x-5}$

Because the domain restriction of $q^{-1}(x)$ is $x \neq \frac{5}{2}$, the range restriction of $q(x)$ is $y \neq \frac{5}{2}$.

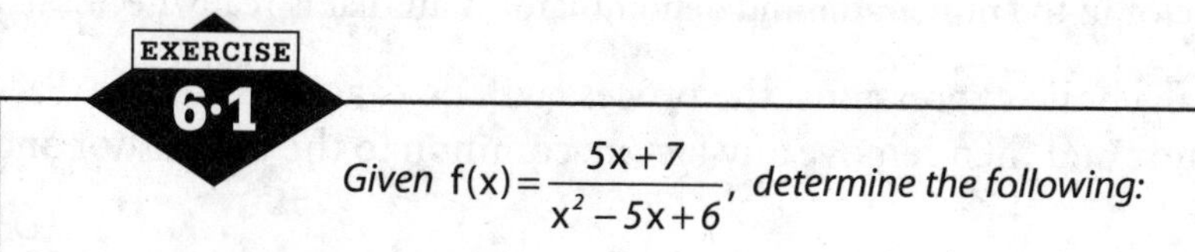

EXERCISE 6·1

Given $f(x) = \dfrac{5x+7}{x^2 - 5x + 6}$, *determine the following:*

1. The domain restriction(s) of $f(x)$
2. The zeroes of $f(x)$
3. The end behavior of $f(x)$

Given $g(x) = \dfrac{2x^2 + 5x - 12}{3x^2 + 19x - 14}$, *determine the following:*

4. The domain restriction(s) of $g(x)$
5. The zeroes of $g(x)$
6. The end behavior of $g(x)$

Given $k(x) = \dfrac{9x+8}{4x-3}$, *determine the following:*

7. The domain restriction(s) of $k(x)$
8. The zeroes of $k(x)$
9. The end behavior of $k(x)$
10. $k^{-1}(x)$

Multiplying and dividing rational expressions

Multiplying rational expressions involves eliminating common factors from the numerator and denominator of the constituent expressions, and then multiplying all remaining factors of the numerator and all remaining factors of the denominator.

PROBLEM Multiply $\dfrac{150}{245} \times \dfrac{252}{280}$ without the use of a calculator.

SOLUTION Factor each of the terms into its prime factors: $150 = 15 \times 10 = 3 \times 5 \times 5 \times 2 = 2 \times 3 \times 5^2$; $245 = 5 \times 49 = 5 \times 7^2$; $252 = 2 \times 126 = 2 \times 2 \times 63 = 2^2 \times 7 \times 9 = 2^2 \times 3^2 \times 7$; $280 = 10 \times 28 = 2 \times 5 \times 4 \times 7 = 2^3 \times 5 \times 7$.

Substitute this factorization for the given terms:

$$\frac{150}{245} \times \frac{252}{280} = \frac{2 \times 3 \times 5^2}{5 \times 7^2} \times \frac{2^2 \times 3^2 \times 7}{2^3 \times 5 \times 7}$$

Combine numerators and denominators: $\dfrac{2^3\times3^3\times5^2\times7}{2^3\times5^2\times7^3}$

Reduce by division of common factors: $\dfrac{3^3}{7^2}=\dfrac{27}{49}$.

Experienced people will not necessarily break these numbers into their prime factors, but the example does help illustrate the basic process. It is not necessary to rewrite $\dfrac{2\times3\times5^2}{5\times7^2}\times\dfrac{2^2\times3^2\times7}{2^3\times5\times7}$ as a single fraction in order to reduce it. Division of a common factor in the numerator and denominator is all that is really necessary.

When multiplying rational algebraic expressions, the process will be exactly the same. Factor the terms into their prime factors and then remove any factors common to the numerator and denominator.

PROBLEM Multiply $\left(\dfrac{6x^2-11x-10}{12x^2+11x+2}\right)\left(\dfrac{12x^2-5x-2}{8x^2-14x-15}\right)$.

SOLUTION Factor each of the terms:

$$\left(\frac{6x^2-11x-10}{12x^2+11x+2}\right)\left(\frac{12x^2-5x-2}{8x^2-14x-15}\right)=\frac{(3x+2)(2x-5)}{(3x+2)(4x+1)}\times\frac{(4x+1)(3x-2)}{(2x-5)(4x+3)}$$

$3x+2$, $4x+1$, and $2x-5$ are factors common to both the numerator and denominator. After these terms are removed,

$$\left(\frac{6x^2-11x-10}{12x^2+11x+2}\right)\left(\frac{12x^2-5x-2}{8x^2-14x-15}\right)=\frac{3x-2}{4x+3}.$$

Note: The domain restrictions of the original problem and the reduced problem are the same. That is, the domain restrictions of the simplified answer are still $x\neq\frac{-2}{3},\frac{-1}{4},\frac{5}{2},\frac{-3}{4}$. The importance of looking for the domain restrictions of the original problem will become clearer when solving rational equations.

Since division is the inverse operation for multiplication, the process of dividing rational algebraic expressions is to change the division to multiplication, and change the divisor to its reciprocal. The problem then becomes a multiplication problem.

PROBLEM Simplify $\dfrac{4x^2-9}{6x^2-x-12}\div\dfrac{8x^2+2x-15}{6x^2+23x+20}$.

SOLUTION Change the division to multiplication and take the reciprocal of the divisor to rewrite the problem as: $\dfrac{4x^2-9}{6x^2-x-12}\times\dfrac{6x^2+23x+20}{8x^2+2x-15}$

Factor each of the terms: $\dfrac{(2x-3)(2x+3)}{(2x-3)(3x+4)}\times\dfrac{(3x+4)(2x+5)}{(2x+3)(4x-5)}$

$2x-3$, $2x+3$, and $3x+4$ are common factors to the numerator and denominator so, $\dfrac{4x^2-9}{6x^2-x-12}\div\dfrac{8x^2+2x-15}{6x^2+23x+20}=\dfrac{2x+5}{4x-5}$.

An important rule for you to remember is that expressions of the form $\dfrac{a-b}{b-a}$ are equal to -1. (If you are unsure of this, pick any two values for a and b and work out the problem.)

PROBLEM Simplify $\frac{10x^2+x-3}{4-8x}\times\frac{8x^2-7x-1}{4x^2+8x-12}\div\frac{x^3-1}{8x^2+8x+8}$.

SOLUTION Change the division to multiplication:

$$\frac{10x^2+x-3}{4-8x}\times\frac{8x^2-7x-1}{4x^2+8x-12}\times\frac{8x^2+8x+8}{x^3-1}$$

Factor the terms: $\frac{(2x-1)(5x+3)}{4(1-2x)}\times\frac{(8x+1)(x-1)}{4(x-1)(x+3)}\times\frac{8(x^2+x+1)}{(x-1)(x^2+x+1)}$

$2x-1$ and $1-2x$ reduce to -1. $\frac{-(5x+3)}{4}\times\frac{(8x+1)(x-1)}{4(x-1)(x+3)}\times\frac{8(x^2+x+1)}{(x-1)(x^2+x+1)}$.

8, x^2+x+1, and $x-1$ are common factors of the numerator and denominator.

Therefore, $\frac{10x^2+x-3}{4-8x}\times\frac{8x^2-7x-1}{4x^2+8x-12}\div\frac{x^3-1}{8x^2+8x+8}=\frac{-(5x+3)(8x+1)}{2(x-1)(x+3)}$.

EXERCISE

6·2

Simplify the following.

1. $\left(\frac{20x^2+7x-3}{16x^2-1}\right)\left(\frac{15x^2+x-6}{25x^2-9}\right)$

2. $\left(\frac{8x^2+29x-12}{12x^2-8x-15}\right)\left(\frac{18x^2-9x-20}{3x^2+8x-16}\right)$

3. $\left(\frac{72x-192}{12x+18}\right)\left(\frac{2x^2+x-3}{40-15x}\right)$

4. $\frac{9x^2-4}{6x^2-13x+6}\div\frac{3x^2-x-2}{x^3-1}$

5. $\frac{100x^2-81}{100x^2+180x+81}\div\frac{20x^2-8x-9}{30x^2-23x-45}$

6. $\frac{x^3+64}{x^2-16}\times\frac{45-60x}{5x^2-20x+80}\div\frac{16x^2-24x+9}{4x^2+13x-12}$

Adding and subtracting rational expressions

Just as adding and subtracting algebraic expressions require "like" terms, so do adding and subtracting rational expressions. While you can add $3x$ and $5x$ to make a single term, $8x$, $3x + 5y$ is as simple as this particular expression can be. When adding $\frac{1}{8}+\frac{5}{8}$, you are adding the number of eighths. The purpose of getting a common denominator is so that you can add or subtract like terms.

PROBLEM Simplify $\frac{5}{x^2-1}+\frac{4}{x^2-2x+1}$.

SOLUTION These denominators are not the same, but they are factorable.

Rewrite the problem with the denominators factored: $\frac{5}{(x-1)(x+1)}+\frac{4}{(x-1)^2}$

The first denominator lacks a second factor of $x-1$, while the second denominator lacks a factor of $x+1$. Introduce these factors into the problem by multiplying each of the fractions by a form of 1:

$$\left(\frac{5}{(x-1)(x+1)}\right)\left(\frac{x-1}{x-1}\right)+\left(\frac{4}{(x-1)^2}\right)\left(\frac{x+1}{x+1}\right)=\frac{5(x-1)}{(x-1)^2(x+1)}+\frac{4(x+1)}{(x-1)^2(x+1)}$$

Now that the denominators are the same (i.e., like terms are available), combine the numerators:

$$\frac{5x-5+4x+4}{(x-1)^2(x+1)}=\frac{9x-1}{(x-1)^2(x+1)}$$

PROBLEM Simplify $\frac{3x+2}{8x^2+10x-3}+\frac{2x-1}{8x^2+14x+3}$.

SOLUTION Factor the denominators: $\frac{3x+2}{(2x+3)(4x-1)}+\frac{2x-1}{(2x+3)(4x+1)}$

The first denominator lacks $4x+1$, while the second denominator lacks $4x-1$.

Get equivalent fractions: $\left(\frac{3x+2}{(2x+3)(4x-1)}\right)\left(\frac{4x+1}{4x+1}\right)+\left(\frac{2x-1}{(2x+3)(4x+1)}\right)\left(\frac{4x-1}{4x-1}\right)$

Combine the numerators: $\frac{(3x+2)(4x+1)+(2x-1)(4x-1)}{(2x+3)(4x-1)(4x+1)}=$

$$\frac{12x^2+11x+2+8x^2-6x+1}{(2x+3)(4x-1)(4x+1)}=\frac{20x^2+5x+3}{(2x+3)(4x-1)(4x+1)}$$

There is no need to expand the denominator into a polynomial. You should check to see if the numerator of the rational expression has a common factor with the denominator. If so, the fraction could be reduced. A common factor between numerator and denominator does not exist for this example.

PROBLEM Simplify $\frac{x-2}{x-1}-\frac{x-4}{3x-3}$.

SOLUTION Factor the denominator: $\frac{x-2}{x-1}-\frac{x-4}{3(x-1)}$

Get equivalent fractions: $\left(\frac{x-2}{x-1}\right)\left(\frac{3}{3}\right)-\frac{x-4}{3(x-1)}$

When combining the numerators, you must be aware of the subtraction symbol in this problem and the fact that it will need to be distributed throughout the second numerator. The use of parentheses in the numerator is important.

$$\frac{3(x-2)-(x-4)}{3(x-1)}=\frac{3x-6-x+4}{3(x-1)}=\frac{2x-2}{3(x-1)}$$

Factor the numerator and reduce the fraction: $\frac{2(x-1)}{3(x-1)}=\frac{2}{3}$

PROBLEM Simplify $\frac{x+2}{2x^2+x-1}-\frac{x-2}{4x^2-1}$.

SOLUTION Factor the denominators: $\frac{x+2}{(2x-1)(x+1)}-\frac{x-2}{(2x-1)(2x+1)}$

Get equivalent fractions: $\left(\frac{x+2}{(2x-1)(x+1)}\right)\left(\frac{2x+1}{2x+1}\right)-\left(\frac{x-2}{(2x-1)(2x+1)}\right)\left(\frac{x+1}{x+1}\right)$

$$\frac{(x+2)(2x+1)-(x-2)(x+1)}{(2x-1)(x+1)(2x+1)}=\frac{2x^2+5x+2-(x^2-x-2)}{(2x-1)(x+1)(2x+1)}$$

$$=\frac{2x^2+5x+2-x^2+x+2}{(2x-1)(x+1)(2x+1)}=\frac{x^2+6x+4}{(2x-1)(x+1)(2x+1)}$$

A complex fraction is a fraction in which the numerator and/or the denominator contain fractions. To change a complex fraction to a simple fraction, multiply the numerator and denominator by the common denominator of the component fractions.

PROBLEM Simplify $\dfrac{\frac{2}{3}+\frac{3}{8}}{\frac{11}{12}-\frac{1}{4}}$.

SOLUTION The common denominator for the component fractions is 24:

$$\left(\frac{\frac{2}{3}+\frac{3}{8}}{\frac{11}{12}-\frac{1}{4}}\right)\left(\frac{24}{24}\right)=\frac{\left(\frac{2}{3}+\frac{3}{8}\right)24}{\left(\frac{11}{12}-\frac{1}{4}\right)24}=\frac{\left(\frac{2}{3}\right)24+\left(\frac{3}{8}\right)24}{\left(\frac{11}{12}\right)24-\left(\frac{1}{4}\right)24}=\frac{16+9}{22-6}=\frac{25}{16}$$

PROBLEM Simplify $\dfrac{\frac{x}{x+1}+\frac{x}{x+2}}{1-\frac{2}{x^2+3x+2}}$.

SOLUTION The common denominator for the component fractions is $(x+1)(x+2)$.

$$\left(\frac{\frac{x}{x+1}+\frac{x}{x+2}}{1-\frac{2}{x^2+3x+2}}\right)\left(\frac{(x+1)(x+2)}{(x+1)(x+2)}\right)=\frac{\left(\frac{x}{x+1}\right)(x+1)(x+2)+\left(\frac{x}{x+2}\right)(x+1)(x+2)}{1(x+1)(x+2)-\left(\frac{2}{x^2+3x+2}\right)(x+1)(x+2)}$$

$$=\frac{x(x+2)+x(x+1)}{x^2+3x+2-2}=\frac{x^2+2x+x^2+x}{x^2+3x}=\frac{2x^2+3x}{x^2+3x}=\frac{x(2x+3)}{x(x+3)}=\frac{2x+3}{x+3}$$

EXERCISE 6·3

Simplify the following.

1. $\dfrac{2x^2-7x-15}{6x^2+x-2}\times\dfrac{2x^2+15x-8}{2x^2+19x+24}$

2. $\dfrac{x^3-27}{2x^2+x-21}\times\dfrac{2x^2+17x+35}{5x^3+15x^2+45x}$

3. $\dfrac{4x^2-5x-6}{16x^2-9}\div\dfrac{2x^2-x-6}{12x^2-x-6}$

4. $\dfrac{81x^2-180x+100}{81x^2-100}\div\dfrac{108x-120}{54x^2+60x}$

5. $\dfrac{x+2}{2x-10}+\dfrac{x+3}{x^2-25}$

6. $\dfrac{3x+1}{2x^2-x-6}+\dfrac{2x-1}{2x^2-3x-9}$

7. $\dfrac{2x+5}{30x-6}-\dfrac{x+7}{25x^2-1}$

8. $\dfrac{3x-4}{6x^2+x-1}-\dfrac{2x+3}{6x^2+5x+1}$

9. $\dfrac{\frac{x+1}{x-1}-\frac{x-1}{x+1}}{\frac{x+3}{x+1}+\frac{x-3}{x-1}}$

10. $\dfrac{\frac{2x+1}{3x-1}+\frac{3x+1}{x+1}}{2-\frac{1+12x-5x^2}{3x^2+2x-1}}$

Solving rational equations

The basic principle for solving rational equations is to multiply by a common denominator to remove the fraction from the problem. Pay careful attention to the domain restrictions of the rational expressions and compare your solutions against these values.

PROBLEM Solve $\frac{5x-1}{x+3}=\frac{7x+10}{2x+5}$.

SOLUTION The common denominator for these fractions is $(x+3)(2x+5)$. Multiply both sides of the equation by this amount to get $(5x-1)(2x+5)=(7x+10)(x+3)$. (You will most likely say this is just cross-multiplication, but you should know that, in reality, you are applying the multiplication property of equality.)

Expand: $10x^2+23x-5=7x^2+31x+30$

Gather terms: $3x^2-8x-35=0$

Factor and solve: $(3x+7)(x-5)=0$

$$x=5,\frac{-7}{3}$$

The domain restrictions of the original problem are $x\neq -3,\frac{-5}{2}$. Neither of these match the solution, so both answers, 5 and $\frac{-7}{3}$, are acceptable.

PROBLEM Solve $\frac{x+8}{x-1}+\frac{3x+4}{x+4}=6$.

SOLUTION The common denominator for the rational expressions is $(x-1)(x+4)$. Multiply both sides of the equation by this expression.

$$\left(\frac{x+8}{x-1}+\frac{3x+4}{x+4}\right)(x-1)(x+4)=(6)(x-1)(x+4)$$

Distribute: $\left(\frac{x+8}{x-1}\right)(x-1)(x+4)+\left(\frac{3x+4}{x+4}\right)(x-1)(x+4)=6(x^2+3x-4)$

Multiply: $(x+8)(x+4)+(3x+4)(x-1)=6x^2+18x-24$

Expand: $x^2+12x+32+3x^2+x-4=6x^2+18x-24$

Gather like terms: $4x^2+13x+28=6x^2+18x-24$

$$0=2x^2+5x-52$$

$$(2x+13)(x-4)=0$$

$$x=\frac{-13}{2},4$$

Neither of these answers disagree with the domain restrictions of $x\neq 1,-4$.

PROBLEM Solve: $\frac{3n+8}{n-2}-\frac{n+5}{n-1}=\frac{42}{n^2-3n+2}$.

SOLUTION The denominator on the right-hand side of the equation factors to become $(n-1)(n-2)$, and is the common denominator for all the fractions. Multiply both sides of the equation by this expression.

$$\left(\frac{3n+8}{n-2}-\frac{n+5}{n-1}\right)(n-1)(n-2)=\left(\frac{42}{n^2-3n+2}\right)(n-1)(n-2)$$

Distribute: $\left(\frac{3n+8}{n-2}\right)(n-1)(n-2)-\left(\frac{n+5}{n-1}\right)(n-1)(n-2)=42$

Distribute again: $(3n+8)(n-1)-(n+5)(n-2)=42$

Expand: $(3n^2+5n-8)-(n^2+3n-10)=42$

Subtract: $3n^2+5n-8-n^2-3n+10=42$

$$2n^2+2n+2=42$$

$$2n^2+2n-40=0$$

Factor: $2(n+5)(n-4)=0$

Solve: $n=-5, 4$

These solutions do not contradict the domain restrictions of the original problem, $n \neq 1, 2$.

PROBLEM Solve $\frac{x}{x+2}+\frac{2}{x-6}=\frac{5x-14}{x^2-4x-12}$.

SOLUTION The denominator on the right factors to $(x+2)(x-6)$, so this serves as the common denominator. Multiply both sides of the equation by this amount.

$$\left(\frac{x}{x+2}\right)(x+2)(x-6)+\left(\frac{2}{x-6}\right)(x+2)(x-6)=\left(\frac{5x-14}{x^2-4x-12}\right)(x+2)(x-6)$$

Distribute: $x(x-6)+2(x+2)=5x-14$

$$x^2-6x+2x+4=5x-14$$
$$x^2-4x+4=5x-14$$
$$x^2-9x+18=0$$
$$(x-6)(x-3)=0$$
$$x=6, 3$$

Because the domain restrictions of the original problem are $x \neq -2, 6$, the solution 6 must be rejected. The answer to this problem is $x=3$. In this example, 6 is called an **extraneous solution**—it is extra and erroneous (wrong).

PROBLEM The total resistance, R_T, in a parallel circuit is computed by the formula $\frac{1}{R_T}=\frac{1}{R_1}+\frac{1}{R_2}+\frac{1}{R_3}+\ldots$ for as many resistors as appear within the circuit. Two resistors with rating r and $r+40$ ohms are placed in a parallel circuit and yield a total resistance of 15 ohms. Determine the rating for each resistor.

SOLUTION Using the equation from above, $\frac{1}{r}+\frac{1}{r+40}=\frac{1}{15}$.

Multiply by the common denominator: $\left(\frac{1}{r}+\frac{1}{r+40}\right)15r(r+40)=\left(\frac{1}{15}\right)15r(r+40)$

Distribute: $15(r+40)+15r=r(r+40)$

$$15r+600+15r=r^2+40r$$

Gather terms: $0=r^2+10r-600$

Factor: $(r+30)(r-20)=0$

$$r=-30, 20$$

The resistor cannot have a negative rating so $r=-30$ is rejected as a solution. With $r=20$, the resistors have ratings of 20 and 60 ohms, respectively.

EXERCISE

6·4

Solve the following.

1. $\frac{c+2}{c-5}=\frac{3c-2}{c-4}$

2. $\frac{90}{c-9}-\frac{48}{c-3}=3$

3. $\frac{n+5}{n^2+n-2}+\frac{n+8}{n-1}=\frac{27}{n+2}$

4. $\frac{a+5}{a-1}+\frac{a+5}{2a+1}=\frac{5a^2+6a+4}{2a^2-a-1}$

5. $\frac{x+5}{x^2+x-2}+\frac{x+8}{x^2+3x-4}=\frac{48}{x^2+6x+8}$

6. $\frac{x+1}{x-3}-\frac{x+2}{x-8}=\frac{10-7x-x^2}{x^2-11x+24}$

7. Capacitance is the measure of a capacitor's ability to hold a charge. If a set of capacitors are put in a circuit in series, the total equivalent capacitance, C_{eq}, is given by the formula $\frac{1}{C_{eq}}=\frac{1}{C_1}+\frac{1}{C_2}+\frac{1}{C_3}+\ldots$ for as many capacitors as are in the circuit. If three capacitors with ratings c farads, $c+20$ farads, and $3c$ farads are placed in a series, the net capacitance is 20 farads. Find the value of each capacitor.

Irrational functions

As discussed in Chapter 5, functions of the form $f(x)=\sqrt[n]{x}$ are the inverse of the functions $g(x)=x^n$. When n is a positive even integer, the domain of $f(x)$ is $x \ge 0$, and the range of $f(x)$ is $y \ge 0$. When n is a positive odd integer, the domain and range of $f(x)$ are the set of real numbers.

PROBLEM Determine the domain and range of the function $f(x)=4\sqrt{2x-3}-1$.

SOLUTION The domain of this function requires that $2x-3 \ge 0$, so $x \ge \frac{3}{2}$. The graph of $f(x)$ is a transformation of the graph of $y=\sqrt{x}$. With regard to the range, the graph of $\sqrt{x}$ is dilated from the x-axis by a factor of 4 and translated down one unit. Dilating from the x-axis has no impact on the range (going from $y \ge 0$ to $4y \ge 0$), but the translation changes the range to $y \ge -1$.

PROBLEM Determine the domain and range of the function $g(x)=\sqrt{x^2-4x-5}$.

SOLUTION Realize that $g(x)$ is a composition of the function $y=\sqrt{x}$ and $y=x^2-4x-5$. The domain of $g(x)$ requires that $x^2-4x-5 \ge 0$, or that $(x-5)(x+1) \ge 0$. Because the graph of $y=x^2-4x-5$ is a concave up parabola with x-intercepts at $x=-1, 5$, the parabola will be at or above the x-axis when $x \le -1$ or $x \ge 5$. Therefore, the domain of $g(x)$ is $x \le -1$ or $x \ge 5$. The range of $g(x)$ is determined by the square root of all the output values given by the parabola when $x \le -1$ or $x \ge 5$. Therefore, the range of $g(x)$ is $y \ge 0$.

PROBLEM Find the domain and range of the function $p(x)=\sqrt{25-4x^2}$.

SOLUTION The parabola $y=25-4x^2$ is concave down with a vertex at (0, 25) and x-intercepts at $x=\pm\frac{5}{2}$. Therefore, the domain of $p(x)$ is $-\frac{5}{2} \le x \le \frac{5}{2}$. The range of the parabola over this interval is $0 \le y \le 25$, so the range of $p(x)$ is the square root of these values, $0 \le y \le 5$.

PROBLEM Determine the domain and range of the function $k(x)=\sqrt{32-12x-2x^2}$.

SOLUTION As before, $y=-2x^2-12x+32$ is a concave down parabola. The vertex of this parabola has coordinates (–3, 50) and x-intercepts at 2 and –8. The domain of $k(x)$ is $-8 \le x \le 2$, while the range is $0 \le y \le 5\sqrt{2}$.

EXERCISE 6·5

Given $f(x)=2\sqrt{3x-4}+5$, *determine the following:*

1. The domain of $f(x)$
2. The range of $f(x)$

Given $g(x)=7-5\sqrt{2-3x}$, *determine the following:*

3. The domain of $g(x)$
4. The range of $g(x)$

Given $k(x)=\sqrt{x^2-12x+27}$, *determine the following:*

5. The domain of $k(x)$
6. The range of $k(x)$

Given $m(x)=\sqrt{32-4x-x^2}$, *determine the following:*

7. The domain of $m(x)$
8. The range of $m(x)$

Simplifying irrational expressions

Some basic rules of working with irrational expressions are perfect powers larger than 1 cannot be kept within the radical; radicands should not contain fractions; and it is common practice to rationalize the denominators of fractions.

PROBLEM Simplify $\sqrt{50}$.

SOLUTION Because $50 = 25 \times 2$, $\sqrt{50}=\sqrt{25}\sqrt{2}=5\sqrt{2}$.

PROBLEM Simplify $\sqrt{9000}$.

SOLUTION $9{,}000 = 900 \times 10$, so $\sqrt{9000}=\sqrt{900}\sqrt{10}=30\sqrt{10}$.

PROBLEM Simplify $\sqrt[3]{72}$.

SOLUTION Because this is a cubed root, you need to look for factors of 72 that are perfect cubes. Since $72 = 8 \times 9$, $\sqrt[3]{72}=\sqrt[3]{8\times 9}=\sqrt[3]{8}\sqrt[3]{9}=2\sqrt[3]{9}$.

PROBLEM Simplify $3\sqrt{20}+4\sqrt{45}$.

SOLUTION Simplify each of the expressions, $3\sqrt{4}\sqrt{5}+4\sqrt{9}\sqrt{5}=6\sqrt{5}+12\sqrt{5}=18\sqrt{5}$.

PROBLEM Simplify $4\sqrt[3]{81}+5\sqrt[3]{192}$.

SOLUTION $\sqrt[3]{81}=\sqrt[3]{27}\sqrt[3]{3}=3\sqrt[3]{3}$ and $\sqrt[3]{192}=\sqrt[3]{64}\sqrt[3]{3}=4\sqrt[3]{3}$.
$4\sqrt[3]{81}+5\sqrt[3]{192}=4(3\sqrt[3]{3})+5(4\sqrt[3]{3})=12\sqrt[3]{3}+20\sqrt[3]{3}=32\sqrt[3]{3}$.

PROBLEM Rationalize the denominator of the fraction $\frac{8}{\sqrt{12}}$.

SOLUTION Multiply both the numerator and the denominator of the fraction by $\sqrt{3}$:
$\left(\frac{8}{\sqrt{12}}\right)\left(\frac{\sqrt{3}}{\sqrt{3}}\right)=\frac{8\sqrt{3}}{\sqrt{36}}$. This becomes $\frac{8\sqrt{3}}{6}=\frac{4\sqrt{3}}{3}$.

PROBLEM Rationalize the denominator of the fraction $\frac{9}{2+\sqrt{3}}$.

SOLUTION Take advantage of the difference of squares product, $(a+b)(a-b)=a^2-b^2$, to remove the radical from the denominator. ($a+b$ and $a-b$ are called conjugates.) Multiply the numerator and the denominator of the fraction by $2-\sqrt{3}$:
$\left(\frac{9}{2+\sqrt{3}}\right)\left(\frac{2-\sqrt{3}}{2-\sqrt{3}}\right)=\frac{9(2-\sqrt{3})}{2^2-(\sqrt{3})^2}=\frac{9(2-\sqrt{3})}{4-3}=9(2-\sqrt{3})$.

EXERCISE 6·6

Simply the expressions in questions 1–4.

1. $\sqrt{175}$
2. $5\sqrt{32}-2\sqrt{50}$
3. $\sqrt[3]{2000}$
4. $\frac{12\sqrt[3]{54}}{\sqrt[3]{432}}$

Rationalize the denominator for each fraction given in questions 5–7.

5. $\frac{10}{\sqrt{75}}$
6. $\frac{26}{4+\sqrt{3}}$
7. $\frac{5+\sqrt{2}}{3+\sqrt{2}}$

Solving irrational equations

The basic format for solving irrational equations is to isolate the radical expression on one side of the equation, and then raise both sides of the resulting equation to an appropriate power to remove the radical.

PROBLEM Solve $\sqrt{3x-2}=5$.

SOLUTION Square both sides of the equation to get $3x-2=25$. Add 2 and divide by 3 to get $x=9$.

PROBLEM Solve $\sqrt[3]{2x-1}-4=3$.

SOLUTION Add 4 to both sides of the equation: $\sqrt[3]{2x-1}=7$
Cube both sides of the equation: $2x-1=343$
Add 1 and divide by 2: $x=172$

PROBLEM Solve $\sqrt{2x+3}+x=16$.

SOLUTION Subtract x from both sides of the equation: $\sqrt{2x+3}=16-x$
Square both sides of the equation: $2x+3=(16-x)^2$
Expand the right-hand side of the equation: $2x+3=x^2-32x+256$
Set the left-hand side to 0: $0=x^2-34x+253$
Factor and solve (or use the quadratic formula): $(x-11)(x-23)=0$
$x=11, 23$

As with rational equations, when solving irrational equations one must be aware of the domain of the original equation to ensure that extraneous roots are not reported as answers. Check each answer for accuracy.

$x=11$: $\sqrt{2(11)+3}+11=16$ becomes $\sqrt{25}+11=16$, which simplifies to $5+11=16$. $x=11$ is a solution.

$x=23$: $\sqrt{2(23)+3}+23=16$ becomes $\sqrt{49}+23=16$, which simplifies to $7+23=16$, a false statement. $x=23$ is not part of the solution.

The solution to this problem is $x=11$.

PROBLEM Solve $\sqrt{3x-2}+\sqrt{4x+1}=9$.

SOLUTION There are two radical expressions in this problem.

Isolate one of the expressions: $\sqrt{3x-2}=9-\sqrt{4x+1}$
Square both sides of the equation: $3x-2=(9-\sqrt{4x+1})^2$
Expand the right-hand side: $3x-2=81-18\sqrt{4x+1}+4x+1$

This is similar to problems seen earlier with only one radical expression.

Isolate the radical: $3x-2=4x+82-18\sqrt{4x+1}$

$18\sqrt{4x+1}=x+84$

Square both sides of the equation: $(18\sqrt{4x+1})^2=(x+84)^2$

Solve: $324(4x+1)=x^2+168x+7{,}056$

$1296x+324=x^2+168x+7{,}056$
$0=x^2-1128x+6732$

Use the quadratic formula to determine that $x=6, 1122$.

Check $x=6$: $\sqrt{3(6)-2}+\sqrt{4(6)+1}=9$ becomes $\sqrt{16}+\sqrt{25}=9$. $4+5=9$ is true.
Check $x=1122$: $\sqrt{3(1122)-2}+\sqrt{4(1122)+1}=9$ becomes $\sqrt{3364}+\sqrt{4489}=9$, which clearly is not true.

Therefore, $x=6$ is the solution to this problem.

EXERCISE
6·7

Solve the following.

1. $\sqrt{5x-9}=11$
2. $\sqrt{19-2x}=7$
3. $\sqrt{3x+22}+x=16$
4. $\sqrt{5x+49}-x=11$
5. $\sqrt{5x+4}+\sqrt{4x+1}=15$
6. $\sqrt{5x+6}-\sqrt{40-x}=4$

Exponential and logarithmic functions

Compound interest, population growth, radioactive decay, the pH of a solution, and the Richter scale are examples of exponential and logarithmic functions. In this chapter, you will review the properties of exponents and the graphs of exponential functions. A logarithmic function will be introduced as the inverse of its exponential function, and this chapter will explore that relationship.

Properties of exponents

You learned about exponents early in your study of mathematics. At that time, you learned that x^2 means "x times x," x^3 means "x times x times x," and so on. Eventually, you came to understand that x^n means to use x as a factor n times. You then should have learned the rules for combining terms with exponents.

RULES OF EXPONENTS

Rule I	$b^m b^n = b^{m+n}$	When multiplying terms, add exponents.
Rule II	$\frac{b^m}{b^n} = b^{m-n}$	When dividing terms, subtract exponents.
Rule III	$(b^m)^n = b^{mn}$	When raising a power to a power, multiply exponents.
Rule IV	$b^0 = 1$	Any number (other than 0) raised to the 0 power is 1.
Rule V	$(ab)^n = a^n b^n$	If a product is raised to a power, the result is the same as the product of the powers.

PROBLEM Simplify each of the following.

a. $\left(\frac{b^7}{b^2}\right)^3$

b. $(4x^3)^3$

c. $\frac{(6x^4)^2}{(3x^2)^4}$

SOLUTION a. $\frac{b^7}{b^2} = b^5$, so $\left(\frac{b^7}{b^2}\right)^3 = (b^5)^3 = b^{15}$

b. $(4x^3)^3 = (4)^3(x^3)^3 = 64x^9$

c. $\frac{(6x^4)^2}{(3x^2)^4} = \frac{(6)^2(x^4)^2}{(3)^4(x^2)^4} = \frac{36x^8}{81x^8} = \frac{4}{9}$

There are two important consequences to these rules, which, although they might not seem intuitive, do follow from the rules of exponents.

The first consequence has to do with negative exponents. Logically, we know that $\frac{b^3}{b^7}$ is equal to $\frac{1}{b^4}$ because there are more factors of b in the denominator of the fraction than there are in the numerator. According to Rule II, we are supposed to subtract the exponent of the denominator from the exponent of the numerator. If we do that, we get b^{-4} as a result. Clearly, there is only one answer to this problem, so these two expressions, b^{-4} and $\frac{1}{b^4}$, must be equivalent. Therefore, by definition, $b^{-n} = \frac{1}{b^n}$. In other words, negative exponents mean reciprocals.

$$\text{Rule VI: } b^{-n} = \frac{1}{b^n}$$

PROBLEM Simplify each of the following.

a. $\frac{b^3}{b^{-5}}$

b. $\frac{b^{-4}}{b^{-5}}$

c. $\frac{a^3b^{-2}}{a^{-4}b^5}$

SOLUTION a. Subtract exponents $(3 - (-5))$ to get b^8.

b. Subtract exponents $(-4 - (-5))$ to get $b^1 = b$.

c. Subtract exponents $(3 - (-4))$ for a, and $(-2 - 5)$ for b, to get a^7b^{-7}. If written without negative exponents, this will equal $\frac{a^7}{b^7}$.

The second consequence has to do with fractional exponents. In earlier math courses you learned about roots of numbers, that is, square roots, cube roots, fourth roots, etc.

We know that $\sqrt{9} = 3$, because $3^2 = 9$; $\sqrt[3]{8} = 2$, because $2^3 = 8$; and $\sqrt[4]{10{,}000} = 10$, because $10^4 = 10{,}000$. Consider the following problem:

$$(b^x)^3 = b$$

Rule III tells us the left-hand side of this equation becomes:

$$b^{3x} = b$$

The right-hand side of this problem can be written as:

$$b^{3x} = b^1$$

In order for these two expressions to be equal, it is logical to state that it must be the case that $3x = 1$, so x must be equal to $\frac{1}{3}$. But what does an exponent of $\frac{1}{3}$ mean? Asking this question a different way, "What number cubed gives an answer of b?" Our experience has shown us that the answer to this question is the cubed root of b.

Consequently, we define rational exponents as roots. That is, $\sqrt[n]{b} = b^{\frac{1}{n}}$. Fractional exponents mean roots.

$$\text{Rule VII: } \sqrt[n]{b} = b^{\frac{1}{n}}$$

We can now write $\sqrt{9} = 9^{\frac{1}{2}} = 3$, $\sqrt[3]{8} = 8^{\frac{1}{3}} = 2$, and $\sqrt[4]{10{,}000} = 10{,}000^{\frac{1}{4}} = 10$. What is the value of $27^{\frac{2}{3}}$? We rewrite $27^{\frac{2}{3}} = \left(27^{\frac{1}{3}}\right)^2$. According to Rule VIII, this is equal to $(\sqrt[3]{27})^2 = 3^2 = 9$.

PROBLEM Simplify each of the following.

a. $36^{\frac{3}{2}}$

b. $(27x^6)^{\frac{2}{3}}$

c. $(25x^4)^{\frac{-1}{2}}$

d. $\left(\dfrac{8x^9}{27y^6}\right)^{\frac{-2}{3}}$

SOLUTION a. $36^{\frac{3}{2}} = \left(36^{\frac{1}{2}}\right)^3 = (6)^3 = 216$

b. $(27x^6)^{\frac{2}{3}} = (27)^{\frac{2}{3}}(x^6)^{\frac{2}{3}} = \left[(27)^{\frac{1}{3}}\right]^2\left[(x^6)^{\frac{1}{3}}\right]^2 = (3)^2(x^2)^2 = 9x^4$

c. $(25x^4)^{\frac{-1}{2}} = \dfrac{1}{(25x^4)^{\frac{1}{2}}} = \dfrac{1}{(25)^{\frac{1}{2}}(x^4)^{\frac{1}{2}}} = \dfrac{1}{5x^2}$

d. $\left(\dfrac{8x^9}{27y^6}\right)^{\frac{-2}{3}} = \left(\dfrac{27y^6}{8x^9}\right)^{\frac{2}{3}} = \dfrac{\left[(27)^{\frac{1}{3}}\right]^2\left[(y^6)^{\frac{1}{3}}\right]^2}{\left[(8)^{\frac{1}{3}}\right]^2\left[(x^9)^{\frac{1}{3}}\right]^2} = \dfrac{(3)^2(y^2)^2}{(2)^2(x^3)^2} = \dfrac{9y^4}{4x^6}$

EXERCISE

7·1

Simplify the following statements. Answers should not include negative exponents.

1. $(2x^3y^2z^4)(5x^2y^3z)$
2. $(2x^4)(5x)^3$
3. $(2x)^4(5x)^3$
4. $\dfrac{8x^3}{12x^5}$
5. $\dfrac{18x^3y^{-4}}{27x^5y^{-2}}$
6. $\dfrac{(3x^2)^4}{(9x^4)^2}$
7. $\dfrac{(2x^3)^3(3x^2)^2}{(6x^5)^2}$
8. $81^{\frac{3}{4}}$

9. $\left(\frac{27}{64}\right)^{\frac{-2}{3}}$

10. $\left(\frac{4x^{11}y^{14}}{9x^3y^{16}}\right)^{\frac{1}{2}}$

11. $\left(\frac{1000x^{-4}y^{-2}}{27x^{-10}y^{-8}}\right)^{\frac{-2}{3}}$

12. $5y^{\frac{2}{3}}\left(2y^{\frac{1}{3}}+3y^{\frac{-2}{3}}\right)$

Exponential and logarithmic functions

Functions of the form $f(x)=Ab^x$, with $b>0$ and $b\neq 1$, are called **exponential functions.** (The value of b is not allowed to be 1, because 1 raised to any power is 1 and the function would then be a constant.) The graphs of exponential functions have the following characteristics: the y-intercept is always $(0, A)$; the graph passes through the point $(1, Ab)$; the domain is the set of real numbers; the function is one to one; and the graph has a horizontal asymptote at $y = 0$. (Recall that an asymptote is a line that a function approaches but will never touch. Note: Asymptotes for exponential functions are always horizontal lines.) If $A > 0$, the range of the function is $y > 0$ (the range is $y < 0$ if $A < 0$). When $b > 1$, the graph increases in value from left to right, and if $0 < b < 1$, the graph decreases in value.

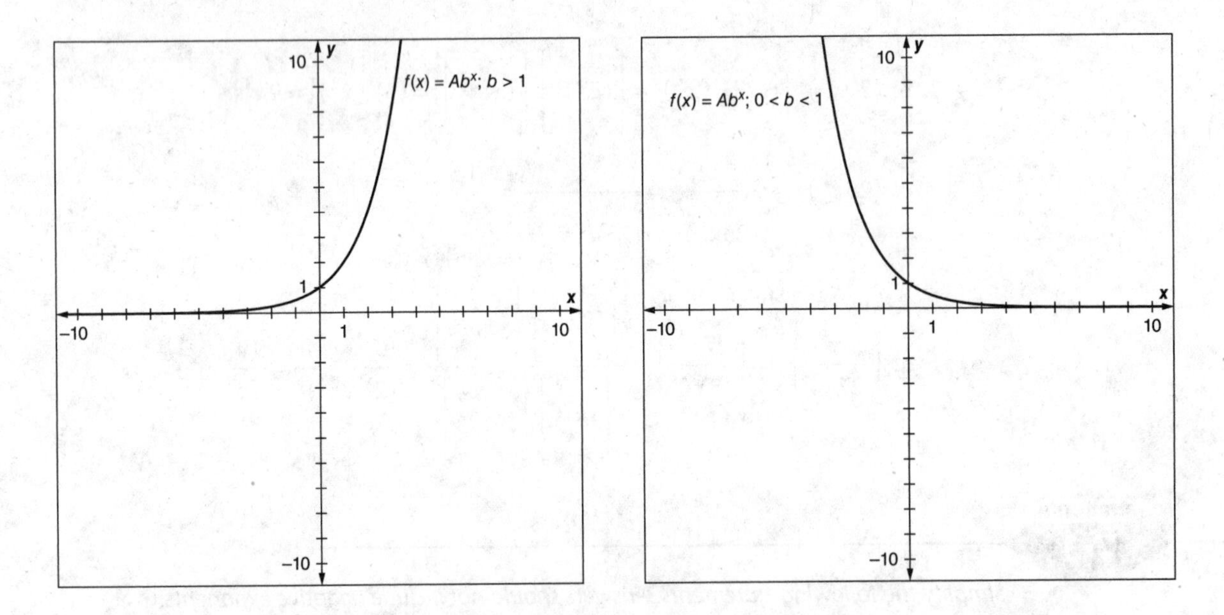

For $b > 1$, as the value of b gets larger, the graph becomes steeper. (You can check this with a graphing utility.) There is a value of the base of the exponential functions that is used more than any other. This value, e, known as **Euler's number** and named after the Swiss mathematician Leonard Euler, is approximately equal to 2.718. Future studies in mathematics will provide you with many examples of its use.

An initial reaction to this number is that it is difficult to understand. It may help you to think of the number in this way: almost everyone above the age of 14 can state that π is approximately 3.14 (or may even be able to go as far as 3.14159), knowing that π is irrational with an infinitely long decimal representation. The number e is just like this. Most scientific and graphing calculators have the exponential functions 10^x and e^x built into their programs with buttons available for easy user access.

PROBLEM Given $f(x) = 4(3^x)$, compute $f(4)$.

SOLUTION $f(4) = 4(3^4) = 4(81) = 324$.

PROBLEM Given $g(x) = e^x$, compute $g(4)$.

SOLUTION $g(4) = e^4$ (which is the exact answer) and is approximately equal to 54.5982.

PROBLEM Describe the transformation of $y = 2^x$ needed to create the sketch of $p(x) = 3(2^{x+1}) - 5$. Sketch a graph of both functions on the same set of axes.

SOLUTION The graph of $p(x)$ takes the graph of 2^x and translates it to the left 1 unit, stretches the graph from the x-axis by a factor of 3, and then translates the graph down 5 units. The point (0, 1) is transformed to (−1, −2). The point (1, 2) translates to (0, 1), and the horizontal asymptote $y = 0$ moves to $y = -5$.

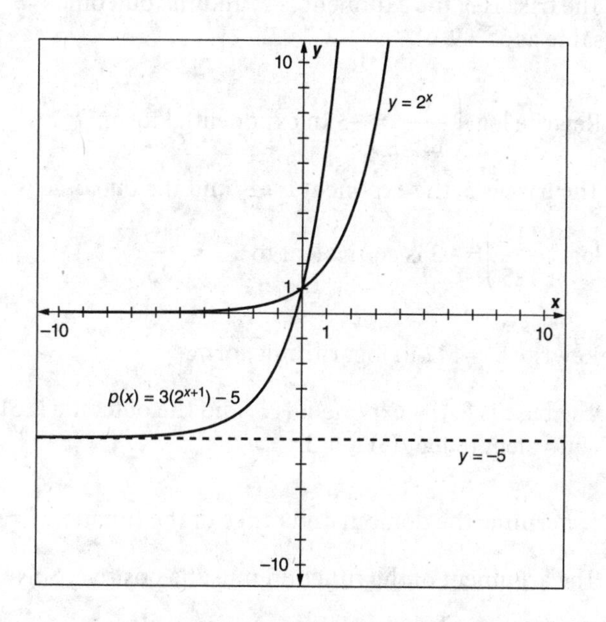

Because the graph of the exponential function $y = b^x$ is one to one, it has an inverse which is a function and has the equation $x = b^y$. A Scottish preacher and astronomer, John Napier, was the first to name this new function. He called it a **logarithmic function** (derived from the Greek for ratio and number). The equation b^y is equivalent to the more common function $y = \log_b(x)$. This notation is read "y = log base b of x." It is critical that you understand that a **logarithm** is an exponent.

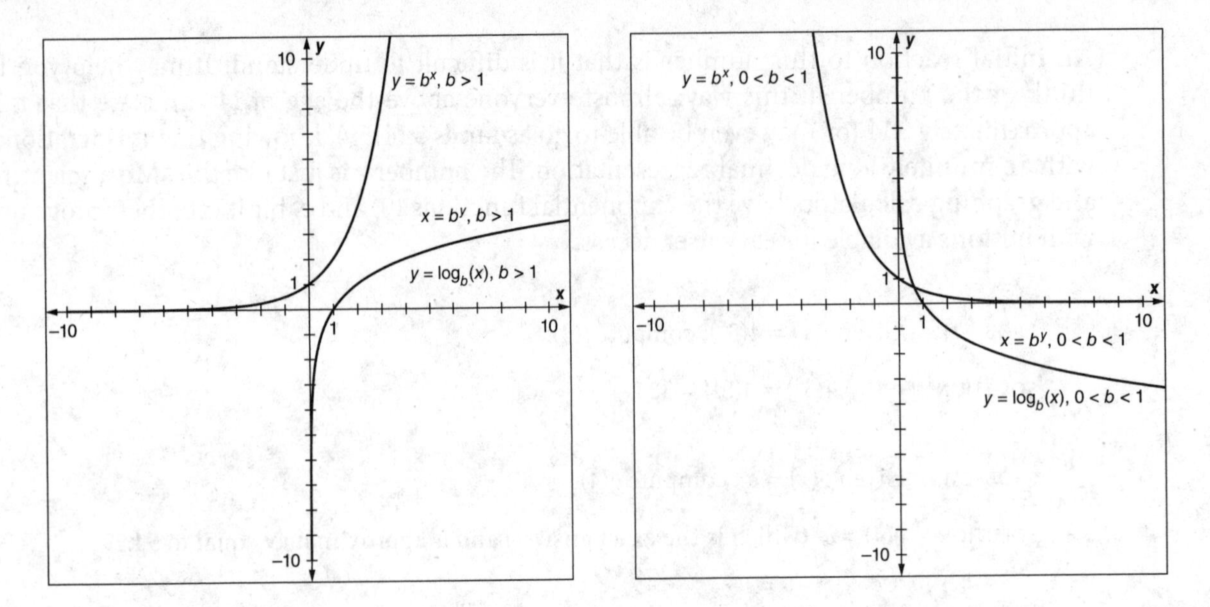

If the domain of the exponential function $y = b^x$ is the set of real numbers, and the range is $y > 0$, then the domain of the function $y = \log_b(x)$ is $x > 0$, and the range is also the set of real numbers. If the asymptotes of exponential functions are horizontal lines in the form $y = 0$, then asymptotes of logarithmic functions are vertical lines in the form $x = 0$.

PROBLEM Rewrite $\log_3(9) = 2$ in exponential form.

SOLUTION The base is 3, the exponent is 2, and the outcome is 9. Therefore, $\log_3(9) = 2$ is the same as $3^2 = 9$.

PROBLEM Rewrite $\log_5\left(\frac{1}{125}\right) = -3$ in exponential form.

SOLUTION The base is 5, the exponent is -3, and the outcome is $\frac{1}{125}$. Therefore, $\log_5\left(\frac{1}{125}\right) = -3$ is equivalent to $5^{-3} = \frac{1}{125}$.

PROBLEM Rewrite $8^3 = 512$ in logarithmic form.

SOLUTION The base is 8, the exponent is 3, and the outcome is 512. Therefore, $8^3 = 512$ is equivalent to $\log_8(512) = 3$.

PROBLEM Determine the domain and range of the function $y = \log_6(2x - 3) - 1$.

SOLUTION The argument of the function must be positive. Solve $2x - 3 > 0$ to get $x > \frac{3}{2}$.

Translating the outcomes down 1 unit does not change the range as the set of real numbers.

As noted earlier, the two most common exponential functions are $y = 10^x$ and $y = e^x$. The inverses of these functions are written in a special manner. The inverse of $y = 10^x$ is $y = \log(x)$, and the inverse of $y = e^x$ is $y = \ln(x)$. $y = \log(x)$ is called the **common log** (because 10 is the common base used universally), and $y = \ln(x)$ is called the **natural logarithm** (because e is the common base).

"ln" is an abbreviation for *logarithmus naturalis*. In the Romance languages, which include Latin, the noun is written before the adjective. Scholarly works throughout much of European history were written in Latin, and many mathematical terms derive from these early works.

PROBLEM Rewrite $10^3 = 1000$ in logarithmic form.

SOLUTION The base is 10, the exponent is 3, and the outcome is 1000. $10^3 = 1000$ is equivalent to $\log_{10}(1000) = 3$, but is commonly written as $\log(1000) = 3$.

PROBLEM Rewrite $e^a = c$ in logarithmic form.

SOLUTION The base is e, the exponent is a, and the outcome is c. $e^a = c$ is equivalent to $\log_e(c) = a$, but is more commonly written as $\ln(c) = a$.

EXERCISE 7·2

Rewrite the equations in questions 1 and 2 in logarithmic form.

1. $9^4 = 6561$
2. $\left(\frac{2}{3}\right)^{-4} = \frac{81}{16}$

Rewrite the equations given in questions 3 and 4 in exponential form.

3. $\log(100) = 2$
4. $\log_3\left(\frac{1}{27}\right) = -3$

Complete questions 5 and 6 as directed.

5. Find the domain and range of $f(x) = 3\log(4-5x)+8$.
6. Describe the transformation of the graph of $y = 3^x$ needed to sketch the graph of $g(x) = 2(3^{x-4}) + 1$.

Properties of logarithms

Each of the rules for exponents that we discussed earlier has a corresponding logarithmic rule.

	PROPERTIES OF EXPONENTS	PROPERTIES OF LOGARITHMS
I	$b^m b^n = b^{m+n}$	$\log_b(MN) = \log_b(M) + \log_b(N)$
II	$\frac{b^m}{b^n} = b^{m-n}$	$\log_b\left(\frac{M}{N}\right) = \log_b(M) - \log_b(N)$
III	$(b^m)^n = b^{mn}$	$\log_b(M^N) = N\log_b(M)$
IV	$b^0 = 1$	$\log_b(1) = 0$
V	$(ab)^n = a^n b^n$	$\log_b((MN)^p) = p\log_b(M) + p\log_b(N)$
VI	$b^{-n} = \frac{1}{b^n}$	$\log_b\left(\frac{1}{M^N}\right) = -N\log_b(M)$
VII	$b^{\frac{1}{n}} = \sqrt[n]{b}$	$\log_b(\sqrt[N]{M}) = \frac{1}{N}\log_b(M)$

PROBLEM Use the properties of logarithms to simplify $\log_b((x^2-4)(x^2+9))$ into logarithms of its factors.

SOLUTION Factor x^2-4 to be $(x+2)(x-2)$. The problem is now to simplify $\log_b((x+2)(x-2)(x^2+9))$. Use Property I, which indicates that the logarithm of a product is the sum of the logarithms, to get $\log_b(x+2)+\log_b(x-2)+\log_b(x^2+9)$.

PROBLEM Use the properties of logarithms to simplify $\log_b\left(\frac{(x+3)^2}{(x+6)(x-7)}\right)$.

SOLUTION Use Property II, which indicates that the logarithm of a quotient is the difference of the logarithms, to get $\log_b((x+3)^2)-\log_b((x+6)(x-7))$. Use Property III, which indicates that the logarithm of a term raised to a power is the product of the powers, to get $2\log_b(x+3)-\log_b((x+6)(x-7))$. Apply Property I to get $2\log_b(x+3)-(\log_b(x+6)+\log_b(x-7))$. Notice that both logarithmic terms are within a set of parentheses because they are both divisors of $(x+3)^2$ in the original fraction. Finally, distribute the minus sign to get $\log_b\left(\frac{(x+3)^2}{(x+6)(x-7)}\right)=$ $2\log_b(x+3)-\log_b(x+6)-\log_b(x-7)$.

PROBLEM Given $\log_b(5)=x$, $\log_b(3)=y$, and $\log_b(2)=z$, express each of the following in terms of x, y, z, and constants:

a. $\log_b(30)$

b. $\log_b(100)$

c. $\log_b(0.6)$

SOLUTION a. $30=2\times3\times5$, so $\log_b(30)=\log_b(2\times3\times5)=\log_b(2)+\log_b(3)+\log_b(5)=$ $z+y+x$.

b. $100=2^2\times5^2$, so $\log_b(100)=\log_b(2^2\times5^2)=2\log_b(2)+2\log_b(5)=2z+2x$.

c. $0.6=\frac{3}{5}$, so $\log_b(0.6)=\log_b\left(\frac{3}{5}\right)=\log_b(3)-\log_b(5)=y-x$.

PROBLEM Rewrite $3\log_b(x+7)+\frac{1}{2}\log_b(x+3)-\log_b(x-3)-3\log_b(x+1)$ as a single logarithm.

SOLUTION Apply Property III to those terms with coefficients:

$$\log_b((x+7)^3)+\log_b\left((x+3)^{\frac{1}{2}}\right)-\log_b(x-3)-\log_b((x+1)^3)$$

Factor the negative from the last terms:

$$\log_b((x+7)^3)+\log_b\left((x+3)^{\frac{1}{2}}\right)-(\log_b(x-3)+\log_b((x+1)^3))$$

Use the product rule as well as the rule for fractional exponents:

$$\log_b((x+7)^3\sqrt{x+3})-\log_b((x-3)(x+1)^3)$$

Use the quotient rule: $\log_b\left(\frac{(x+7)^3\sqrt{x+3}}{(x-3)(x+1)^3}\right)$

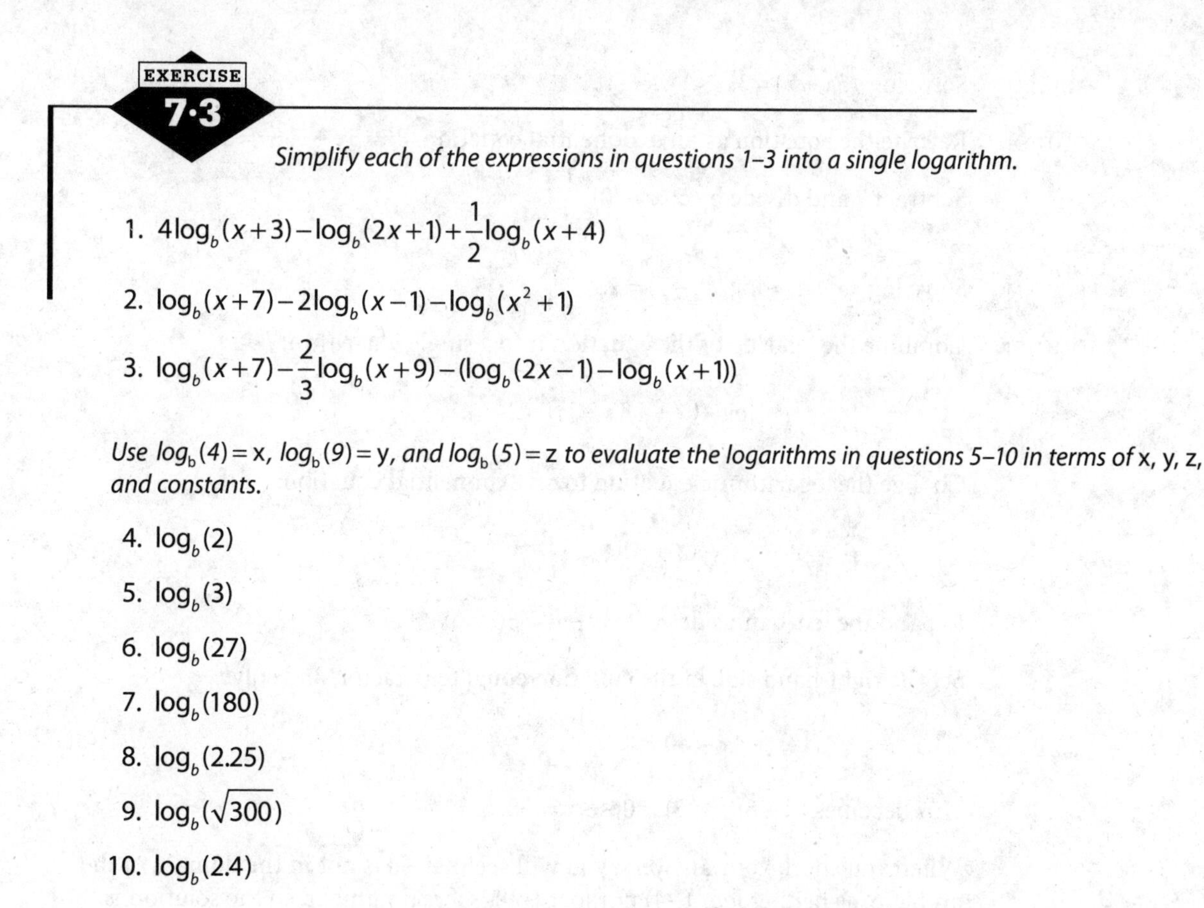

EXERCISE

7·3

Simplify each of the expressions in questions 1–3 into a single logarithm.

1. $4\log_b(x+3)-\log_b(2x+1)+\frac{1}{2}\log_b(x+4)$
2. $\log_b(x+7)-2\log_b(x-1)-\log_b(x^2+1)$
3. $\log_b(x+7)-\frac{2}{3}\log_b(x+9)-(\log_b(2x-1)-\log_b(x+1))$

Use $\log_b(4)=x$, $\log_b(9)=y$, and $\log_b(5)=z$ to evaluate the logarithms in questions 5–10 in terms of x, y, z, and constants.

4. $\log_b(2)$
5. $\log_b(3)$
6. $\log_b(27)$
7. $\log_b(180)$
8. $\log_b(2.25)$
9. $\log_b(\sqrt{300})$
10. $\log_b(2.4)$

Solving exponential and logarithmic equations

The basic process for solving exponential equations is to rewrite both sides of the equation with a common base and use the principle if $b^x = b^y$, then $x = y$. The solution to logarithmic equations starts with the conversion of the logarithmic equation to its equivalent exponential equation.

PROBLEM Solve $3^x = 27$.

SOLUTION Rewrite 27 as 3^3. The problem is now $3^x = 3^3$, so $x = 3$.

PROBLEM Solve $9^x = 27$.

SOLUTION $9 = 3^2$, so rewrite the equation as $(3^2)^x = 3^3$ or $3^{2x} = 3^3$. Set $2x = 3$, which means $x = \frac{3}{2}$.

PROBLEM Solve $8^{2x+3} = 4^{4x-5}$.

SOLUTION Both 8 and 4 are powers of 2.

Rewrite the equation: $(2^3)^{2x+3} = (2^2)^{4x-5}$

Simplify the exponents: $(2)^{6x+9} = (2)^{8x-10}$

Set the exponents equal to each other: $6x + 9 = 8x - 10$

This becomes $2x = 19$ or $x = 9\frac{1}{2}$.

PROBLEM Solve $\log_8(5x+7)=3$.

SOLUTION Rewrite the equation as an exponential equation: $8^3 = 5x + 7$. $8^3 = 512$

Subtract 7 and divide by 5: $x = 101$

PROBLEM Solve $\log_6(x+4)+\log_6(x-1)=2$.

SOLUTION Combine the left side of the equation into a single logarithm:

$$\log_6((x+4)(x-1))=2$$

Change the logarithmic equation to an exponential equation:

$$(x+4)(x-1)=6^2$$

Expand the left-hand side: $x^2 + 3x - 4 = 36$

Set the right-hand side of the equation equal to 0, factor, and solve:

$$x^2 + 3x - 40 = 0$$

This becomes $(x + 8)(x - 5) = 0$, so $x = -8, 5$.

When you check your answers, you will see that -8 is not in the domain of the problem, as neither $\log_6(-4)$ nor $\log_6(-9)$ is a real number, so the solution is $x = 5$.

Some problems do not lend themselves to common bases, so the use of the logarithm function on a calculator is needed.

PROBLEM Solve $8^x = 150$.

SOLUTION 8 and 150 do not have a common base. To solve this equation, take the logarithm of both sides (it does not matter if you use the common logarithm or natural logarithm). $\log(8^x) = \log(150)$. Apply the power rule to the left-hand side of the equation to get $x\log(8)$. Divide by $\log(8)$ to get $x = \frac{\log(150)}{\log(8)} = 2.4096$ (rounded to 4 decimal places). Although $\log(150)$ and $\ln(150)$ have different values, as do $\log(8)$ and $\ln(8)$, $\frac{\log(150)}{\log(8)}$ and $\frac{\ln(150)}{\ln(8)}$ are equal.

PROBLEM Solve $12(2.3)^x = 450$.

SOLUTION Divide each side of the equation by 12: $2.3^x = \frac{450}{12}$

Take the logarithm of both sides of the equation and apply the power rule to the left side of the equation:

$$x\log(2.3)=\log\left(\frac{450}{12}\right)$$

Divide by log(2.3): $x = \dfrac{\log\left(\dfrac{450}{12}\right)}{\log(2.3)}$

Enter this expression into your calculator to get $x = 4.3514$ (rounded to four decimal places).

PROBLEM Solve $28e^{-0.034x} = 23$.

SOLUTION Divide by 28: $e^{-0.034x} = \dfrac{23}{28}$

Take the natural logarithm of both sides of the equation: $\ln(e^{-0.034x}) = \ln\left(\dfrac{23}{28}\right)$

Apply the power rule to the left side of the equation: $-0.034x\ln(e) = \ln\left(\dfrac{23}{28}\right)$

Because $\ln(e) = 1$ (Property IV), divide by -0.034 to solve:

$x = \dfrac{\ln\left(\dfrac{23}{28}\right)}{-0.034} = 5.7856$ (correct to four decimal places).

PROBLEM Evaluate $\log_{12}(93)$.

SOLUTION Set the expression equal to x and rewrite the equation in exponential form: $12^x = 93$

Take the logarithm of both sides: $\log(12^x) = \log(93)$

Apply the power rule for logarithms: $x\log(12) = \log(93)$

Divide: $x = \dfrac{\log(93)}{\log(12)} = 1.8241$ (rounded to four decimal places)

It is important to note that this problem is a single example of the more general problem, $\log_b(x) = \dfrac{\log(x)}{\log(b)}$. This equation is called the **change of base equation** and works for all legitimate values of b and x.

PROBLEM In 1997, the population of the United States was 270 million people, and the annual growth rate was 1.1%. (Assume that $t = 0$ equates to January 1, 1997.)

a. Based on the model, what was the population of the United States on January 1, 2000?

b. Based on the model, when will the population reach 300 million?

SOLUTION The population each year will be 1.011 times the population from the previous year (100% + 1.1%). Consequently, the population of the United States t years after 1997 is given by $P(t) = 270\,(1.011)^t$, where P is measured in millions.

a. 2000 is 3 years after 1997: $P(3) = 270(1.011)^3 = 279.008$

b. $P = 300$ yields the equation: $300 = 270(1.011)^t$

Divide by 270: $\dfrac{300}{270} = (1.011)^t$

Take the logarithm of both sides of the equation:

$$\log\left(\frac{300}{270}\right) = t\log(1.011)$$

Solve for t: $$\frac{\log\left(\frac{300}{270}\right)}{\log(1.011)} = t = 9.6308$$

$t = 9.6308$ corresponds to 7 months $(.6308 \times 12 = 7.6)$ into the year 2006. The model predicts the U.S. population would reach 300 million in July of 2006. (The population actually reached 300 million on October 17, 2006.)

PROBLEM Xiao deposits $400 into an account that pays 4.2% interest, compounded quarterly. How much time is needed before he will have $600 in his account?

SOLUTION The formula for compound interest is $A = P\left(1+\frac{r}{n}\right)^{nt}$, where P is the amount of principal, r is the annual percentage rate, n is the number of annual interest periods, and t is the number of years. Using this formula, the equation for our problem is:

$$600 = 400\left(1 + \frac{.042}{4}\right)^{4t}$$

$$600 = 400(1.0105)^{4t}$$

$$\frac{600}{400} = (1.0105)^{4t}$$

$$1.5 = (1.0105)^{4t}$$

$$\log(1.5) = \log((1.0105)^{4t})$$

$$\log(1.5) = 4t\log(1.0105)$$

$$\frac{\log(1.5)}{\log(1.0105)} = 4t$$

$$t = \frac{1}{4}\frac{\log(1.5)}{\log(1.0105)} = 9.704$$

Since money is being compounded quarterly, Xiao will have $600 in 9.75 years.

PROBLEM As you might well imagine, Xiao does not want to wait 9.75 years. He would like to have the money in 3 years. What annual rate of interest, compounded quarterly, will he need to achieve this?

SOLUTION Fitting the data into the formula, we get the equation:

$$600 = 400\left(1+\frac{r}{4}\right)^{12}$$

$$1.5 = \left(1+\frac{r}{4}\right)^{12}$$

If we raise both sides of the equation to the $\frac{1}{12}$ power, we get:

$$(1.5)^{\frac{1}{12}} = \left(\left(1+\frac{r}{4}\right)^{12}\right)^{\frac{1}{12}}$$

$$(1.5)^{\frac{1}{12}} = 1+\frac{r}{4}$$

$$1.0343661 = 1+\frac{r}{4}$$

$$0.0343661 = \frac{r}{4}$$

$$r = .13746$$

Xiao will need to invest his money at an annual rate of 13.746% in order to have $600 after 3 years. Notice that we did not need to use logarithms to solve this equation. This is because the variable was in the base of the equation, not the exponent.

PROBLEM The Richter scale is a system to register the intensity of earthquakes. A Richter scale value is the logarithm of the ratio of the vibrations of the earth during the earthquake to the normal vibrations. That is, the Richter value is found from the formula:

$$R = \log\frac{\text{Intensity}_{\text{quake}}}{\text{Intensity}_{\text{non-quake}}}$$

The normal rate at which the earth vibrates differs from locale to locale.

The December 26, 2004, earthquake that caused the tsunamis that devastated Thailand and Sri Lanka was first thought to have a Richter value of 9. Later studies indicated a value of 9.3. (The largest recorded Richter value, 9.5, was during the 1960 earthquake in Chile.)

a. The original reading of 9 means that this section of the Indian Ocean sea floor was vibrating 10^9 (1,000,000,000) times more than normal. What does a Richter value of 9.3 mean?

b. How many times more violent is an earthquake with a Richter value of 9.3 than an earthquake with a Richter value of 9?

c. How many times more violent is an earthquake with a Richter value of 6.3 than an earthquake with a Richter value of 6?

SOLUTION a. A Richter value of 9.3 means that the earth was vibrating $10^{9.3} = 1.99526 \times 10^9$ times its normal rate.

b. An earthquake with a Richter value of 9.3 vibrates $10^{0.3} = 1.99526$ times as much as an earthquake with a Richter value of 9.

c. An earthquake with a Richter value of 6.3 vibrates $10^{0.3} = 1.99526$ times as much as an earthquake with a Richter value of 6.

EXERCISE

7·4

Solve the following equations.

1. $25^{x+4} = 125^{3-x}$
2. $4^{3x+1} = (\sqrt{2})^{5x+8}$
3. $\log_{12}(x^2 - 25) = 2$
4. $\log(x) + \log(x - 3) = 1$

Solve the following equations. Round your answers to the nearest hundredth.

5. $12e^{2.3x} = 807$
6. $82(4.3^{5.1x}) = 23{,}415$

pH values are examples of logarithmic functions. pH measures the concentration of the hydrogen ion in a solution, [H+]. pH is computed as the negative logarithm of [H+]: pH = – log([H+]). A solution with a high concentration of [H+] is called an acid, while a solution with a low concentration of [H+] is called a base. Use this information to answer the following.

7. Determine the [H+] for each of the following:
   a. Lemon juice: pH = 2.4
   b. Sea water: pH = 8.0
8. Pure water, with [H+] = 0.0000001, is called a neutral solution. Which of the items in question 7 is considered an acid and which is a base?

Sequences and series

·8·

Late in the eighteenth century, a five-year-old student and his German grammar school classmates were asked to find the sum of the first 100 counting numbers. Before the teacher had a chance to sit down, the five-year-old brought his slate to the teacher with the correct answer on it. The student, Carl F. Gauss (1777–1855), would grow up to be one of the world's most famous mathematicians and physicists. His analysis of the problem was the beginning of the study of arithmetic sequences and series. In this chapter, you will learn more about these topics.

Summation notation

Gauss and his classmates were told to find the sum of the first 100 counting numbers. Before you learn how Gauss approached the problem, you will learn some notation for problems like this.

Restating the problem, find $1+2+3+4+\cdots+97+98+99+100$. Writing all 100 numbers would be very tedious. Mathematicians use the upper-case Greek letter sigma (Σ) to represent **summation**. The notation, $\sum_{\text{first}}^{\text{last}} \text{rule}$, consists of three inputs: the rule that generates the numbers, the first input value, and the last input value. Note: Input values can only be integer values, more specifically, counting or natural numbers. For example, $1+2+3+4+\cdots+97+98+99+100$ can be written more succinctly as $\sum_{n=1}^{100} n$. Start with 1, then go to 2, then 3, and so on, until 100 is reached. Then, add the terms.

PROBLEM Write the expanded form of the summation $\sum_{n=1}^{5}(2n-1)$ and find the sum.

SOLUTION The rule is $2n-1$. When 1 is substituted for n, $2(1)-1=1$. When 2 is substituted for n, $2(2)-1=3$. $n=3$ yields 5, $n=4$ yields 7, and $n=5$ yields 9. Therefore, $\sum_{n=1}^{5}(2n-1)=1+3+5+7+9=25$.

PROBLEM Write the expanded form for $\sum_{n=3}^{7}(n^2-1)$ and find the sum.

SOLUTION Be sure to notice that the first term is $n=3$. When $n=3$, $(3)^2-1=8$. When $n=4$, $(4)^2-1=15$. $n=5$ yields 24, $n=6$ yields 35, and $n=7$ yields 48. Therefore, $\sum_{n=3}^{7}(n^2-1)=8+15+24+35+48=130$.

PROBLEM Write the sum $2+4+6+8+\cdots+40$ using sigma notation.

SOLUTION All the numbers are even. Even numbers, by definition, must be of the form $2n$ (where n is an integer). There are 20 numbers in this summation so $\sum_{n=1}^{20}(2n)$ will represent the problem.

PROBLEM Rewrite $4+7+10+13+\cdots+52+55$ in summation notation.

SOLUTION The difficult part here is to recognize the pattern. Do you see that each number in the pattern is one more than a multiple of three? This means that each term is of the form $3n+1$, with 4 being the first term $[3(1)+1=4]$, and 55 being the eighteenth term $[3(18)+1=55]$. Therefore, $4+7+10+13+\cdots+52+55=\sum_{n=1}^{18}(3n+1)$.

EXERCISE

8·1

Expand each of the summation problems in questions 1–4.

1. $\sum_{n=1}^{5}(4n+3)$
2. $\sum_{n=1}^{4}(n^3+1)$
3. $\sum_{n=15}^{18}(5n+10)$
4. $\sum_{n=1}^{4}(3^n+1)$

Rewrite each of the following questions using summation notation.

5. $5+11+17+23+29+35$
6. $120+115+110+105+100$
7. $81+89+97+105+\cdots+153$
8. $2+4+8+16+32+64+\cdots+2048$

Recursion

A **sequence** is a listing of elements. The elements of the list are separated by commas.
Some sequences are more familiar than others.

S, M, T, W, T, F, S is the sequence for the first letters of the days of the week, beginning with Sunday.
1, 2, 3, 4, 5, … is a sequence of the counting numbers.
O, T, T, F, F, S, S, E, N, T is the sequence of the first letter of the first ten counting numbers when spelled out.
1, 1, 2, 3, 5, 8, 13, 21, … is the **Fibonacci sequence**. (This sequence is named for the medieval mathematician, Leonardo Fibonacci.)

If you do not know about the Fibonacci sequence, it would be worth your while to do some research to learn more about Fibonacci—a mathematician of Italian birth who traveled and lived in northern Africa and the Mediterranean region during the twelfth and early thirteenth centuries—and the sequence named for him. Fibonacci promoted the Hindu–Arabic numeral system to Europeans.

1. Think of a number.
2. Add 5 to it.
3. Add 5 to the answer you just got.
4. Repeat step 3 until your answer is greater than 100.

How many terms did it take?

The answer for each step in the problem depends upon the number that came from the previous step. In essence, this is the **recursive process**. You start with a single term, or some fixed number of terms, and all subsequent terms depend upon the initial value(s).

PROBLEM Given $a_0 = 1$ and $a_n = 2a_{n-1}$, write the first five terms for this process.

The variable a is used to represent the terms in the sequence, and the subscript identifies the place of the term in the sequence. That is, a_1 is the first term and a_7 is the seventh term.

SOLUTION The problem states that the initial term is 1. $a_1 = 2a_{1-1} = 2a_0 = 2(1) = 2$. Following this reasoning, $a_2 = 2a_{2-1} = 2a_1 = 2(2) = 4$; $a_3 = 2a_2 = 2(4) = 8$; and, $a_4 = 2a_3 = 2(8) = 16$. The recursive definition provided in the problem statement gives the values for the function $f(x) = 2^x$, provided the domain is the set of whole numbers.

PROBLEM Find the next six terms of the sequence defined recursively by $a_1 = 1$, $a_2 = 1$, and $a_n = a_{n-2} + a_{n-1}$ $(n \geq 2)$.

SOLUTION The third term, a_3, is equal to $a_{3-2} + a_{3-1} = a_1 + a_2 = 1 + 1 = 2$.

$a_4 = a_{4-2} + a_{4-1} = a_2 + a_3 = 1 + 2 = 3$
$a_5 = a_{5-2} + a_{5-1} = a_3 + a_4 = 2 + 3 = 5$
$a_6 = a_{6-2} + a_{6-1} = a_4 + a_3 = 3 + 5 = 8$
$a_7 = a_{7-2} + a_{7-1} = a_5 + a_6 = 5 + 8 = 13$
$a_8 = a_{8-2} + a_{8-1} = a_6 + a_7 = 8 + 13 = 21$

These are the first eight terms of the Fibonacci sequence.

PROBLEM Find the next five terms of the sequence defined recursively by $a_0 = 1$, $a_n = n\ a_{n-1}$.

SOLUTION The next term, a_1, is defined by $a_1 = 1 \times a_0 = 1(1) = 1$.

$a_2 = 2\ a_1 = 2(1) = 2$
$a_3 = 3\ a_2 = 3(2) = 6$
$a_4 = 4\ a_3 = 4(6) = 24$
$a_5 = 5\ a_4 = 5(24) = 120$

The term a_n gives the value n.

PROBLEM Compute $\sum_{n=1}^{5} a_n$, in which $a_1 = 3$ and $a_n = (a_{n-1})^2 - 3a_{n-1} + 1$.

SOLUTION $\sum_{n=1}^{5} a_n = a_1 + a_2 + a_3 + a_4 + a_5$. However, you must go through all the steps of computing a_2, a_3, a_4, and a_5 before you can compute the sum. $a_2 = (3)^2 - 3(3) + 1 = 1$, $a_3 = (1)^2 - 3(1) + 1 = -1$, $a_4 = (-1)^2 - 3(-1) + 1 = 5$, and $a_5 = (5)^2 - 3(5) + 1 = 11$.

Therefore, $\sum_{n=1}^{5} a_n = 3 + 1 - 1 + 5 + 11 = 19$.

PROBLEM Given $a_1 = 8$ and $a_n = 4(a_{n-1})^{-1}$, find the next three terms of the sequence.

SOLUTION $a_2 = 4(8)^{-1} = 4\left(\frac{1}{8}\right) = \frac{1}{2}$. $a_3 = 4\left(\frac{1}{2}\right)^{-1} = 4 \times 2 = 8 \times a_4 = 4(8)^{-1} = 4\left(\frac{1}{8}\right) = \frac{1}{2}$. This sequence alternates between the numbers $\frac{1}{2}$ and 8 for all terms past the first term.

EXERCISE

8·2

Find the next five terms in the sequences given.

1. $a_1 = 7; a_n = a_{n-1} + 9$
2. $a_1 = 78; a_n = a_{n-1} - 6$
3. $a_1 = 4; a_n = 5\, a_{n-1}$
4. $a_1 = 2; a_n = 3\, a_{n-1} + 1$
5. $a_1 = 6; a_2 = 11; a_n = a_{n-1} + a_{n-2}$
6. $a_1 = 4; a_n = 8\,(a_{n-1})^{-2}$
7. Find $\sum_{n=1}^{4} a_n$ if $a_1 = 2; a_n = (n-1)\, a_{n-1}$

Arithmetic sequences

An **arithmetic sequence** is one in which the difference between consecutive terms is a constant. Letting d represent the common difference, the nth term of an arithmetic sequence can be represented by the formula $a_n = a_1 + d(n-1)$.

PROBLEM Given 12 as the first term of an arithmetic sequence with a common difference of 9, determine the value of the fifteenth term.

SOLUTION $a_{15} = 12 + 9(15 - 1) = 12 + 9(14) = 138$.

PROBLEM Find the twentieth term of the sequence 9, 17, 25, 33, 41, ...

SOLUTION Take the time to check that the difference between terms is a constant. There will be sequences other than arithmetic that you will encounter. The common difference in this example is 8. Therefore, the twentieth term is $a_{20} = 9 + 8(19) = 161$.

PROBLEM Find the eightieth term of an arithmetic sequence, if the twelfth term is 93 and the twentieth term is 149.

SOLUTION The twentieth term, 149, can be written as $149 = a_1 + d(19)$ or $149 = a_1 + 19d$. The twelfth term, 93, can be written as $93 = a_1 + d(11)$ or $93 = a_1 + 11d$.

Recognize that this is a system of equations. Subtract the equations from each other to get $56 = 8d$ or $d = 7$. Substitute 7 for d to determine $a_1 = 16$. $a_{80} = 16 + 7(79) = 569$.

It should be noted that the equation for generating the nth term of an arithmetic sequence is a linear equation. $a_n = a_1 + d(n-1)$ becomes $a_n = a_1 + dn - d$ or $a_n = dn + a_1 - d$. Letting $b = a_1 - d$, the equation becomes $a_n = dn + b$. The common difference is the slope of the line, and the difference between the first term and the common difference is the y-intercept of the line.

Determine the value of the specified term in the arithmetic sequences given.

1. The fortieth term of the arithmetic sequence where 914 is the first term and –7 is the common difference
2. The eightieth term of the arithmetic sequence where 87 is the first term and 19 is the common difference
3. The hundredth term of the sequence 8, 20, 32, 44, ...
4. The seventy-fifth term of the sequence 1023, 1012, 1001, 1090, ...
5. The fifty-seventh term of the arithmetic sequence in which $a_{15} = 89$ and $a_{24} = 224$
6. The hundred-twentieth term of the arithmetic sequence in which $a_5 = 9$ and $a_{34} = 212$

Arithmetic series

In the story told at the beginning of this chapter, young Carl Gauss determined the sum of the first 100 counting numbers. The counting numbers form an arithmetic sequence and the sum of the numbers form an **arithmetic series**. (In general, a **series** is the sum of the terms of a sequence.) How was Gauss able to do the problem so easily? Gauss wrote S as the sum of the numbers, and he then rewrote the sum placing the numbers in reversed order.

$$S = 1 + 2 + 3 + \cdots + 98 + 99 + 100$$

$$S = 100 + 99 + 98 + \cdots + 3 + 2 + 1$$

Adding these equations to each other, he got

$$2S = 101 + 101 + 101 + \cdots + 101 + 101 + 101.$$

The right-hand side of the equation has 101 added 100 times, so $2S = 100(101)$ or $S = \dfrac{100(101)}{2} =$ 5050. Examining this process, you can see that 100 was the number of terms being added and that 101 is the sum of the first term (1) and the last term (100). The denominator, 2, comes from adding the two equations together. The formula for the sum of n terms of an arithmetic series is $S_n = \dfrac{n}{2}(a_1 + a_n)$, or $S_n = \dfrac{n}{2}(a_1 + a_1 + (n-1)d) = \dfrac{n}{2}(2a_1 + (n-1)d)$.

PROBLEM Find the sum of the first 120 terms of the arithmetic sequence, where 19 is the first term and 6 is the common difference.

SOLUTION The last term to be added is $a_{120} = 19 + 6(119) = 733$. Therefore, $S_{120} = \dfrac{120}{2}(19 + 733)$ $= 45{,}120$.

PROBLEM Find the sum of the first 80 terms in the series $7 + 15 + 23 + 31 + \cdots$.

SOLUTION The first term is 7 and the common difference is 8. Therefore, $a_{80} = 7 +$ $8(79) = 639$. $S_{80} = \dfrac{80}{2}(7 + 639) = 25{,}840$. The problem can also be done as $S_{80} = \dfrac{80}{2}(2(7) + 8(79))$ when using the second form of the sum.

PROBLEM Find the sum of the series $34 + 45 + 56 + 67 + \cdots + 683$.

SOLUTION The series is arithmetic, where 34 is the first term and 11 is the common difference. Before the sum can be found, the number of terms in the series must be determined. The nth term of this arithmetic sequence is 683, so $683 = 34 + 11(n - 1)$. Solve this equation to determine $n = 60$.

$$S_{60} = \frac{60}{2}(34 + 683) = 21{,}510$$

PROBLEM The seating in a theater has 58 seats in the first row. Each row after that has one more seat than the row in front of it. If the total number of seats in the theater is 2,175, how many rows of seats does the theater have?

SOLUTION In the language of arithmetic sequences and series, $a_1 = 58$, $d = 1$, and $S_n = 2175$. Using the formula $S_n = \frac{n}{2}(2a_1 + (n-1)d)$, $2175 = \frac{n}{2}(2(58) + (n-1)(1))$. Multiply by 2 to get $4350 = n(115 + n)$. This becomes the quadratic equation, $n^2 + 115n - 4350 = 0$. Factor to get $(n + 145)(n - 30) = 0$, so $n = 30$. (You can also use the quadratic formula to get the same result.)

EXERCISE

8·4

Solve the following.

1. Find the sum of the first 90 terms of an arithmetic series where 350 is the first term and -4 is the common difference.
2. Find the sum of the first 40 terms of an arithmetic series where -226 is the first term and 17 is the common difference.
3. Find S_{60} for the series $21 + 27 + 33 + 39 + \cdots$.
4. Find S_{225} for the series $7 + 11 + 15 + 19 + \cdots$.
5. Find the sum of the series $13 + 27 + 41 + 55 + \cdots + 419$.
6. Find the sum of the series $812 + 799 + 786 + 773 + \cdots + 45$.

Geometric sequences

A **geometric sequence** is one in which the ratio between consecutive terms is a constant. Letting r represent the common ratio, the nth term of a geometric sequence can be represented by the formula $a_n = a_1 \times r^{n-1}$.

PROBLEM Find the eighth term in the geometric sequence where 4 is the first term and 3 is the common ratio.

SOLUTION $a_8 = 4 \times 3^{8-1} = 4 \times 3^7 = 8748$.

PROBLEM Find the twelfth term in the sequence 128,000, 64,000, 32,000, 16,000, … .

SOLUTION The first term is 128,000 and the common ratio is $\frac{1}{2}$. The twelfth term is $a_{12} = 128{,}000\left(\frac{1}{2}\right)^{11} = \frac{125}{2}$.

PROBLEM Find the fifteenth term in a geometric sequence in which $a_3 = 15$ and $a_8 = 480$.

SOLUTION $a_8 = 480$ means that $480 = a_1r^7$, and $a_3 = 15$ means that $15 = a_1r^2$. Divide the first equation by the second to get $32 = r^5$. Solve to get $r = 2$. Substitute $r = 2$ into the second equation to get $a_1 = \frac{15}{4}$. The fifteenth term of the sequence is $a_{15} = \frac{15}{4}(2)^{14} =$ 61,440.

PROBLEM Compound interest is an example of a geometric sequence. Suppose \$5000 is put into an account that pays 3% compounded annually. How much money will be in the account after 10 years? After 20 years? After 30 years?

SOLUTION One year after the money is placed in the account there will be $5000 + 5000(.03) = 5000(1.03)$ dollars in the account—the original \$5000 plus interest earned. The ratio between consecutive terms is 1.03.

Note that the exponent is 1 in this case, not 0. This is because the phrase "one year after" represents the second term of the sequence, with the initial deposit being the first term.

After 10 years, there will be $5000(1.03)^{10} = \$6{,}719.58$.
After 20 years, there will be $5000(1.03)^{20} = \$9{,}030.56$.
After 30 years, there will be $5000(1.03)^{30} = \$12{,}136.31$.

It is worth noting that when terms are in a geometric sequence (e.g., $r, r^2, r^3, r^4, \ldots$) the exponents are in an arithmetic sequence. This is the essence of logarithms.

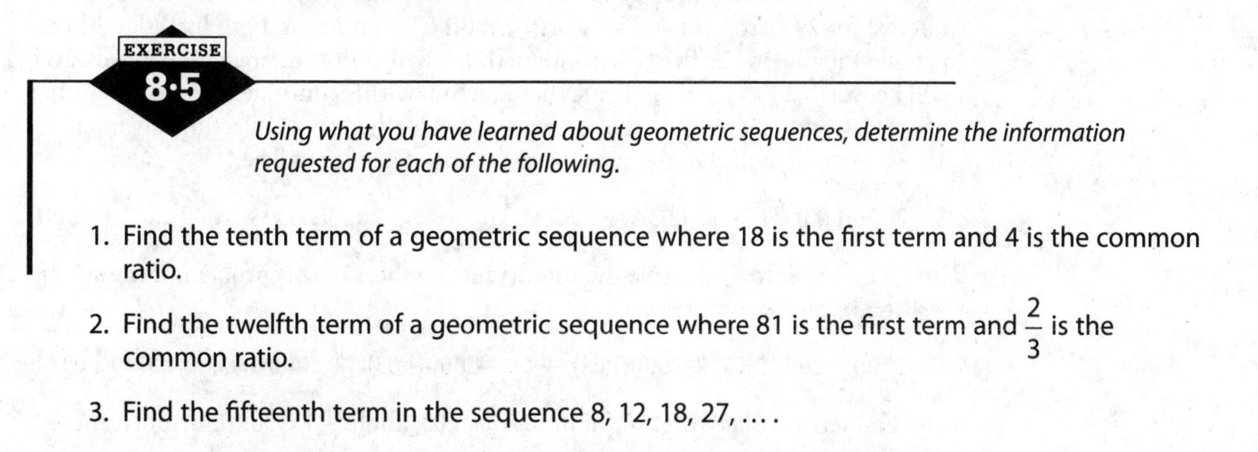

EXERCISE 8·5

Using what you have learned about geometric sequences, determine the information requested for each of the following.

1. Find the tenth term of a geometric sequence where 18 is the first term and 4 is the common ratio.
2. Find the twelfth term of a geometric sequence where 81 is the first term and $\frac{2}{3}$ is the common ratio.
3. Find the fifteenth term in the sequence 8, 12, 18, 27, … .
4. Find the ninth term in the sequence 80,000, 20,000, 5000, 2500, … .
5. Find the twentieth term of a geometric sequence in which the fifth term is 256 and the ninth term is 4096.
6. \$10,000 is invested in an account that pays 2.5% compounded annually. How much money will be in the account after 15 years?

Geometric series

The sum of the terms in a geometric sequence is called a **geometric series.** The derivation for the formula for the sum of a geometric series is different from the way in which Gauss determined the sum of the first 100 counting numbers. Let S_n represent the sum of the first n terms of the geometric series.

$$S_n = a_1 + a_1r + a_1r^2 + a_1r^3 + \cdots + a_1r^{n-1}$$

Multiply both sides of the equation by r, the common ratio.

$$rS_n = a_1r + a_1r^2 + a_1r^3 + \cdots + a_1r^{n-1} + ar^n$$

Notice how all but the first term in the first equation and the last term in the second equation have matches. Subtract the second equation from the first equation.

$$S_n - rS_n = a_1 - a_1r^n$$

Factor and solve for S_n to get $S_n = \dfrac{a_1(1-r^n)}{1-r}$.

PROBLEM Find the sum of the first 20 terms in a geometric series where 5 is the first term and 2 is the common ratio.

SOLUTION $S_{20} = \dfrac{5(1-2^{20})}{1-2} = 5{,}242{,}875.$

PROBLEM Beginning on her thirtieth birthday, Cheryl deposits \$2000 into an account that pays 3% interest compounded annually. Cheryl continues to make \$2000 payments each year on her birthday, with the last payment on her sixtieth birthday. How much money will be in the account after her last deposit?

SOLUTION The first payment will collect interest for 30 years and will be worth $2000(1.03)^{30}$ on her sixtieth birthday. The payment made on her thirty-first birthday will collect interest for 29 years and will be worth $2000(1.03)^{29}$ on her sixtieth birthday. The payment made on her thirty-second birthday will collect interest for 28 years and will be worth $2000(1.03)^{28}$ on her sixtieth birthday. This continues for the remaining payments, with the last payment on her sixtieth birthday not collecting any interest. The amount of money in the account after her last deposit will be:

$$S = 2000(1.03)^{30} + 2000(1.03)^{29} + 2000(1.03)^{28} + \cdots + 2000(1.03)^2 + 2000(1.03) + 2000$$

It will be easier to determine the important numbers of this problem if the sum is written in reverse order:

$$S = 2000 + 2000(1.03) + 2000(1.03)^2 + \cdots + 2000(1.03)^{28} + 2000(1.03)^{29} + 2000(1.03)^{30}$$

The first term is 2000, the common ratio is 1.03, and $n - 1 = 30$ (remember, there are 31 payments being made).

$$S = \frac{2000(1-(1.03)^{31})}{1-1.03} = 100{,}005.36$$

This is an example of a simple annuity problem that people use to create a retirement account. Cheryl deposited a total of \$62,000 over 31 years and gained more than \$38,000 in interest.

PROBLEM Find S_{30} for the geometric series $900+600+400+\cdots$.

SOLUTION The geometric series has a first term of 900 and a common ratio of $\frac{600}{900}=\frac{2}{3}$.

$$S_{30}=\frac{900\left(1-\left(\frac{2}{3}\right)^{30}\right)}{1-\frac{2}{3}}$$, which is approximately 2699.99.

PROBLEM Find S_{100} for the geometric series $900+600+400+\cdots$.

SOLUTION When you type the formula for $S_{100}=\frac{900\left(1-\left(\frac{2}{3}\right)^{100}\right)}{1-\frac{2}{3}}$ into your calculator, the solution will be 2700.

There is an important concept being applied here that must be addressed. Because the powers of $\frac{2}{3}$ get smaller as the exponents increase, there comes a point in the series where the next term added is negligible. If the number of terms is extended infinitely, the sum approaches a constant. That is, $\left(\frac{2}{3}\right)^n$ approaches 0 as n gets infinitely large. However, this statement only applies when the ratio is between –1 and 1. Otherwise, the expression r^n will get infinitely large.

The **sum of an infinite geometric series** with $|r|<1$ is $S_\infty=\frac{a_1}{1-r}$. Note that terms will be written as sums of terms but will end with $+\cdots$ to show that this series extends indefinitely.

PROBLEM Find the sum of $12+6+3+\frac{3}{2}+\cdots$.

SOLUTION The first term is 12 and the common ratio is $\frac{1}{2}$, so $S_\infty=\frac{12}{1-\frac{1}{2}}=24$.

PROBLEM Find the sum of $0.9+0.09+0.009+0.0009+\cdots$.

SOLUTION The first term is 0.9 and the common ratio is 0.1. Therefore, $S_\infty=\frac{0.9}{1-0.1}=\frac{0.9}{0.9}=1$. The repeating decimal $.\overline{9}=1$.

EXERCISE 8·6

Find the sums for the following series.

1. $\sum_{n=1}^{10} 12(4)^{n-1}$
2. The first 20 terms in the series $20+60+180+540+\cdots$
3. $120{,}000+60{,}000+30{,}000+15{,}000+\cdots$
4. $0.72+0.0072+0.000072+0.00000072+\cdots$
5. The first 15 terms of a geometric series if the fourth term is 24 and the ninth term is 768

·9· Introduction to probability

As you have most likely learned in the past, **probability** is the ratio of the number of ways in which a successful outcome can occur to the total number of possible outcomes. Counting the number of ways in which success can occur can be tricky. This chapter begins with counting issues, and works toward computing probabilities of special cases.

Fundamental theorem of counting

A classic example of counting is the multistep problem. The **fundamental theorem of counting** states that the total number of ways in which a task can be completed is equal to the product of the number of ways in which each step in the sequence can be completed. Two examples are ordering a meal from a menu and creating a personal identification number (PIN).

PROBLEM The cafeteria offers its students a hot meal, a drink, and a dessert. There are four hot meals (pizza, soup and grilled cheese sandwich, tacos, or pasta) to choose from, six drinks (whole milk, skim milk, chocolate milk, apple juice, orange juice, or water), and five desserts (apple, orange, brownie, ice cream sandwich, or cookie of the day). How many different meals can be made?

SOLUTION A student who orders pizza and water can order one of the five desserts. That is five different meals for a student who chooses pizza and water. The same is true for the student who chooses pizza and whole milk, pizza and skim milk, pizza and chocolate milk, pizza and apple juice, or pizza and orange juice. This represents 30 different meals for a student who chooses pizza. This thought process can be repeated for the other hot meal choices.

The number of different lunches available at the cafeteria is equal to the product of the number of ways a student can pick a hot meal (4), a drink (6), and a dessert (5). According to the Fundamental Theorem of Counting, this means that there are $4 \times 6 \times 5 = 120$ different meals available.

PROBLEM A bank customer must select a four-digit PIN from the digits 0–9 for her ATM card. How many different PINs are possible?

SOLUTION The customer has 10 choices for the first digit, 10 for the second digit, 10 for the third digit, and 10 for the fourth digit for a total of $10^4 = 10{,}000$ possibilities.

PROBLEM The cafeteria has two specials running today. The first is the usual hot meal (four options), drink (six options), and dessert (five options). The second is a sandwich (ten options), a drink (four options), and a dessert (seven options). How many different lunches are possible?

SOLUTION There are $4 \times 6 \times 5 = 120$ options for the hot meal, and $10 \times 4 \times 7 = 280$ options for the sandwich meal. Because the choice is one or the other, these options are added together to get a total of 400 different lunch options.

PROBLEM Alan, Bob, Colin, Don, and Ed are going to have their picture taken together. They decide that what they would really like to do is to take a picture for each possible arrangement of the five of them. They estimate it will take 15 sec to move from one arrangement to another. How much time will it take for them to get all of the pictures taken?

SOLUTION There are five choices for who can be first in line. The second in line can be chosen from one of the remaining four. There are three choices for third, two choices for fourth, and one choice for last in line. Using the Fundamental Theorem of Counting, this gives $5 \times 4 \times 3 \times 2 \times 1 = 120$ different arrangements. At 15 sec per picture, this is 1800 sec or 30 min to take all the pictures.

The product of the first n counting numbers is called ***n* factorial** and is written as $n!$. The product of the first five counting numbers is $5! = 120$, as was seen in the last example.

PROBLEM Compute 8!

SOLUTION $8! = 8 \times 7 \times 6 \times 5 \times 4 \times 3 \times 2 \times 1 = 40{,}320$

Note that 8! can be written as $8 \times 7!$, $8 \times 7 \times 6!$, $8 \times 7 \times 6 \times 5!$, etc.

PROBLEM Compute $\frac{7!}{4!}$.

SOLUTION $\frac{7!}{4!} = \frac{7 \times 6 \times 5 \times 4!}{4!} = 7 \times 6 \times 5 = 210.$

PROBLEM Compute (5!)(4!).

SOLUTION $5! = 120$ and $4! = 24$. Therefore, $(5!)(4!) = (120)(24) = 2880$.

EXERCISE 9·1

Solve the following problems.

1. A restaurant is offering a special deal for a three-course meal at a fixed price. The meal consists of a salad, an entrée, and a dessert. If the menu contains two salad choices, five entrées, and three desserts, how many different meals can be ordered?

2. "I scream! You scream! We all scream for ice cream!" Patty goes to her favorite ice cream parlor to order a triple-scoop cone with sprinkles. When she gets there, she is pleased to see that the parlor offers 12 different flavors of ice cream. She has a choice of a sugar, wafer, or waffle cone, and they have chocolate as well as rainbow sprinkles. How many different ice cream cones can Patty order?

3. Compute $6! + 4!$

4. Compute $(3!)^2 + (3^2)!$

5. Compute $\frac{10!}{6!}$

6. Compute $\frac{n!}{(n-2)!}$

7. Alan, Bob, Colin, Don, and Ed had such a good time taking their pictures, they talk their friends Frank and George into joining them the next day to have pictures of the seven of them taken together. With 15 sec between pictures, how much time will it take for the seven of them to have their group pictures taken in all possible orders?

Permutations

There are times when the order of arrangement of objects is important, and other times when it is not. A **permutation** is an arrangement of objects in which order matters. A **combination** is an arrangement of objects in which order does not matter.

Order matters when the arrangement represents a code. Words are codes (BEAT and ABET use the same letters but are different words). License plate tags, telephone numbers, PINs, and security access codes are all examples of permutations. (Note: The access code to get into your locker really is your locker permutation, not your locker combination. This is an unfortunate mix-up in the use of the words.)

PROBLEM Alan, Bob, Colin, Don, and Ed decide to have their pictures taken with only three of them in the picture at a time, and they want to be sure to have all possible arrangements. They still want to have all of the possible picture permutations. How many different pictures need to be taken?

SOLUTION There are five choices for the first person, four choices for the second person, and three choices for the third person, or $5 \times 4 \times 3 = 60$ choices altogether.

This product is the same as $\frac{5!}{2!} = \frac{5!}{(5-3)!}$.

PROBLEM Alan, Bob, Colin, Don, and Ed are joined by their friends Frank, George, Hal, Ivan, and José to have pictures taken. They want to have pictures taken four at a time and want to be sure to have all possible arrangements. How many different pictures must be taken?

SOLUTION There are ten choices for the first person, nine choices for the second person, eight choices for the third person, and seven choices for the fourth person. They will need to take $10 \times 9 \times 8 \times 7 = 5040$ pictures. This product is equal to $\frac{10!}{6!} = \frac{10!}{(10-4)!}$.

The two examples hint at the formula for the number of permutations of r objects taken from a group of n objects: ${}_nP_r = \frac{n!}{(n-r)!}$.

PROBLEM If Alan, Bob, Colin, Don, Ed, Frank, George, Hal, Ivan, and José want to have their pictures taken six at a time in all the different arrangements the friends can form, how many different pictures need to be taken?

SOLUTION ${}_{10}P_6 = \frac{10!}{(10-6)!} = \frac{10!}{4!} = \frac{10 \times 9 \times 8 \times 7 \times 6 \times 5 \times 4!}{4!} = 151{,}200$

Most scientific and graphing calculators will have a permutation function. Use it rather than performing this long calculation.

PROBLEM Alan, Bob, Colin, Don, and Ed want to have their pictures taken five at a time in all the different arrangements they can form. How many different pictures need to be taken?

SOLUTION You answered this question in the last section—the answer is 120. What is important in this example is how it appears in the formula. In this problem, the number of ways the five people can be chosen is ${}_5P_5 = \frac{5!}{(5-5)!}$. This is equal to $\frac{5!}{0!} = \frac{120}{0!}$. In order for this expression to equal 120, 0! must be defined to equal 1.

PROBLEM Using all six letters, in how many different ways can the letters in the word ANDREW be arranged?

SOLUTION Using all six letters from the six letters in ANDREW gives ${}_6P_6 = \frac{6!}{(6-6)!} = \frac{6!}{0!} =$ 720, or 720 different arrangements of the letters.

PROBLEM Using all five letters, in how many ways can the letters in the word ALANA be arranged?

SOLUTION This problem is more challenging because there are three As in the problem. For the moment, let's rewrite the word as aL*a*N**a**, with the different formats of the lowercase letter used for emphasis. aL*a*N**a**, **a**L*a*Na, and *a*LaN**a** look like different words when written using these different forms of "a." In fact, for each arrangement of the letters in the word ALANA, interchanging the three As will not create a different word. There are 5! for arranging the 5 letters. However, for each arrangement, there are 3! ways that the three As can be interchanged. The total number of arrangements for the letters in the word ALANA is $\frac{5!}{3!} = 20$.

In general, if a word with n letters has r_1 repetitions of one letter, r_2 repetitions of a second letter, etc., the number of arrangements of the letters is $\frac{n!}{r_1!r_2!\ldots}$.

PROBLEM Using all nine letters, in how many ways can the letters in the word TENNESSEE be arranged?

SOLUTION There are nine letters in the word TENNESSEE, including two Ns, two Ss, and four Es. The number of words that can be made from these letters is $\frac{9!}{2!2!4!} = 3780$.

EXERCISE 9·2

Solve the following.

1. Compute ${}_9P_5$.
2. Compute ${}_{10}P_6$.
3. How many different arrangements can be made using all of the letters in the word KATHRYN?
4. How many different arrangements can be made using all of the letters in the word RUSSELL?
5. The locks on the lockers in the high school have a three-digit code chosen from the 36 numbers on the dial. If the code consists of three different numbers, how many different codes can be made?

6. The positions on a basketball team are assigned numbers to designate their responsibilities (for example, the #1 player is the point guard while #2 is the shooting guard). There are five different positions during a scrimmage, and there are 12 players on the team. If Coach Treanor randomly assigns five of these players to different positions, how many different teams can Coach Treanor put on the floor?
7. Stacey is one of the 12 players on Coach Treanor's team. What is the probability that she is on the team chosen to play?
8. Kieran's Little League team has 15 players on the roster. He puts each name on a card and puts the cards into a hat. His batting order for the day will be the first nine players he picks, with the players batting in the order in which they are picked. What is the probability that Will and Peter, two of his players, will be the first two players to bat?

Combinations

Alice, Barbara, Cathy, Donna, and Edith want to have their pictures taken three at a time. They do not care about who stands next to whom, or the order in which they are standing. How many different pictures need to be taken? You have seen that there were 60 different pictures needed when the guys had their pictures done. How does this number change if order is not a consideration? Suppose Alice, Barbara, and Cathy are the first three to have their pictures taken. If order mattered, ABC, ACB, BAC, BCA, CAB, and CBA (using the first letters of their names) would all be different arrangements. Since order does not count, these 3!(6) arrangements are all the same. Therefore, according to the girls' requirements, there are only $\frac{60}{6} = 10$ pictures that need to be taken.

The number of combinations of n things taken r at a time is ${}_nC_r = \frac{n!}{r!(n-r)!}$. Another notation for ${}_nC_r$ is $\binom{n}{r}$.

PROBLEM Compute $\binom{10}{3}$ and $\binom{10}{7}$.

SOLUTION $\binom{10}{3} = \frac{10!}{3!7!} = 120$, while $\binom{10}{7} = \frac{10!}{7!3!} = 120$.

It will always be the case that $\binom{n}{r}$ and $\binom{n}{n-r}$ will be equal.

PROBLEM Seven seniors and eight juniors are on the homecoming committee. Five students need to be chosen for the float judging committee.

a. How many different combinations could there be for the make-up of the judging committee?
b. How many of the possible combinations will include three seniors and two juniors?
c. How many of the possible combinations will contain more seniors than juniors?
d. What is the probability that the judging committee formed will contain three seniors and two juniors?
e. What is the probability that the judging committee formed will contain more seniors than juniors?
f. Angie is one of the students on the homecoming committee. What is the probability she is on the float judging committee?

SOLUTION a. There are 15 people eligible to be on the committee and there is nothing to indicate that the order of selection matters. Therefore, this is an example of a combination. The number of possible combinations for the judging committee is $\binom{15}{5} = {}_{15}C_5 = \frac{15!}{5!10!} = 3003$.

b. There are $\binom{7}{3} = {}_7C_3 = \frac{7!}{3!4!} = 35$ ways to select three seniors, and $\binom{8}{2} = {}_8C_2 = \frac{8!}{2!6!} = 28$ ways to select two juniors. There are $\binom{7}{3}\binom{8}{2} = 35 \times 28 = 980$ different ways in which the judging committee can be formed.

It is worth noting that the sum of the upper numbers in the combination formula is 15 (the number of eligible students), while the denominators add to 5 (the number of students being selected).

c. A committee with more seniors than juniors will consist of three seniors and two juniors, four seniors and one junior, or five seniors and no juniors. Therefore, the possible number of judging committees with more seniors than juniors is $\binom{7}{3}\binom{8}{2} + \binom{7}{4}\binom{8}{1} + \binom{7}{5}\binom{8}{0} = 980 + (35)(8) + (21)(1) = 1{,}281$ different committees.

d. The probability that the judging committee will contain three seniors and two juniors is $\frac{980}{3003}$.

e. The probability that the judging committee will contain more seniors than juniors is $\frac{1281}{3003}$.

f. There is $\binom{1}{1} = 1$ way to choose Angie. The other four members of the judging committee are drawn from the 14 remaining members of the homecoming committee, $\binom{14}{4} = 1001$. There is a $\frac{1001}{3003} = \frac{1}{3}$ chance that Angie will be on the judging committee.

Pascal's triangle, an interesting item in the study of mathematics, provides a graphical approach for computing combinations. The triangle can be constructed from two different algorithms. The first involves a recursion formula. Each row begins and ends in a 1. All entries between these 1s are the sum of two elements diagonally above it in the triangle.

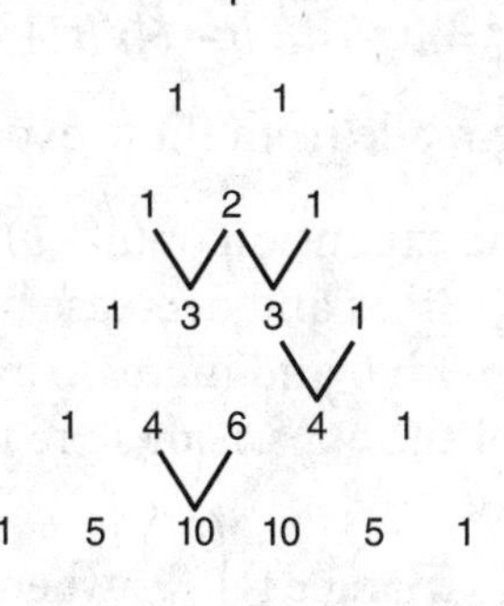

With the top row numbered $n = 0$, the second row numbered $n = 1$, etc., each entry in the triangle can be computed using the combination formula $\binom{n}{r}$, where r starts with 0 and finishes with n. For example, the last line in the diagram shown is $n = 5$. Thus, the numbers for this line are computed as

$$\binom{5}{0} = 1, \binom{5}{1} = 5, \binom{5}{2} = 10, \binom{5}{3} = 10, \binom{5}{4} = 5, \text{ and } \binom{5}{0} = 1.$$

EXERCISE

9·3

Compute the number of possible combinations for the following.

1. Compute ${}_{10}C_6$.

2. Compute ${}_{12}C_8$.

Seven students are to be selected to work on the homecoming committee. There are 12 seniors and 10 juniors to choose from. Given this information, answer the following.

3. How many different possible combinations are there for the committee?
4. What is the probability that the committee formed will contain more seniors than juniors?
5. Marian and Kristen are two of the eligible students. What is the probability that they will both be on the committee?

Eight cards are placed in a box. On each card is written the lengths of the sides of a triangle (3, 4, 5; 5, 12, 13; 1, $\sqrt{3}$, 2; 8, 15, 19; 11, 60, 61; 7, 24, 25; 20, 30, 40; 12, 16, 21). Given this information, solve the following.

6. If four cards are drawn at random from the box, what is the probability that three of them will have lengths that represent a right triangle?

Binomial expansions

The expansions for $(a + b)^n$ are shown for $n = 0, 1, 2, 3,$ and 4.

$$(a+b)^0 = 1$$
$$(a+b)^1 = a + b$$
$$(a+b)^2 = a^2 + 2ab + b^2$$
$$(a+b)^3 = a^3 + 3a^2b + 3ab^2 + b^3$$
$$(a+b)^4 = a^4 + 4a^3b + 6a^2b^2 + 4ab^3 + b^4$$

The following observations can be made from these expansions:

- The number of terms in the expansion of $(a + b)^n$ is $n + 1$.
- The exponents on a begin with n and decrease by 1 to reach 0 (remember $a^0 = 1$).
- The exponents on b begin with 0 and increase by 1 to reach n.
- The coefficients for each of the expansions are the same as seen in Pascal's triangle.

Each coefficient for row n in Pascal's triangle is $\binom{n}{r}$, where r begins with 0 and grows to n (the same as the exponents for b).

PROBLEM Expand $(a+b)^5$.

SOLUTION The coefficients are $\binom{5}{0}=1$, $\binom{5}{1}=5$, $\binom{5}{2}=10$, $\binom{5}{3}=10$, $\binom{5}{4}=5$, and $\binom{5}{5}=1$.

Therefore, $(a+b)^5 = a^5 + 5a^4b + 10a^3b^2 + 10a^2b^3 + 5ab^4 + b^5$.

PROBLEM Expand $(2x-y)^5$.

SOLUTION The result will be the same as in the previous problem, with a replaced by $2x$ and b replaced by $-y$.

$$(2x-y)^5 = (2x)^5 + 5(2x)^4(-y) + 10(2x)^3(-y)^2 + 10(2x)^2(-y)^3 + 5(2x)(-y)^4 + (-y)^5$$
$$(2x-y)^5 = 32x^5 + 5(16x^4)(-y) + 10(8x^3)(y^2) + 10(4x^2)(-y^3) + 5(2x)(y^4) - y^5$$
$$(2x-y)^5 = 32x^5 - 80x^4y + 80x^3y^2 - 40x^2y^3 + 10xy^4 - y^5$$

PROBLEM Find the coefficient of x^6 in the expansion of $(2x+3)^8$.

SOLUTION The term with x^6 in the expansion will be $c(2x)^6(3)^2$, where c represents the binomial coefficient $\binom{8}{r}$. Because the value of r is the same as the exponent of the second term in the binomial, the term with x^6 will be $\binom{8}{2}(2x)^6(3)^2 = 16{,}128x^6$, making the coefficient 16,128.

PROBLEM Find the middle term in the expansion of $\left(4x-\frac{1}{2}\right)^8$.

SOLUTION The expansion will have nine terms, so the middle term will be the fifth term. The exponent for each term will be 4 (half of the exponent 8). The middle term will be $\binom{8}{4}(4x)^4\left(\frac{-1}{2}\right)^4 = 70(256x^4)\left(\frac{1}{16}\right) = 1120x^4$.

EXERCISE 9·4

Expand the following two expressions.

1. $(2x+y)^5$
2. $(x-2y)^6$

Find the specified term in the expansion of the three binomials given below.

3. The fifth term in the expansion of $(3a-4b)^9$
4. The middle term in the expansion of $\left(3x-\frac{2}{3}\right)^{10}$
5. The last term of the expansion of $\left(3x-\frac{2}{3}\right)^{10}$

Find the coefficient of x^4 in the expansion of the following binomial.

6. $(2x-3)^9$

Conditional probability

What is the probability of selecting an 8 from a well-shuffled bridge deck (52 cards in four suits—spades, hearts, clubs, and diamonds)? Since there is one 8 in each of the four suits, the probability of getting an 8 is $P(8) = \frac{4}{52} = \frac{1}{13}$. This type of question places no conditions on the likelihood of achieving the intended outcome. A problem of this type is known as an **unconditional probability** problem.

Using the same well-shuffled deck, what is the probability of selecting an 8 on the second draw, if the first card drawn (and not returned to the deck) was a king? The difference between this question and the one in the previous paragraph is that there is a condition attached: the first card drawn, and not replaced, is a king. The probability of getting an 8 on the second pick, given that a king is drawn on the first pick, is written $P(8|K)$, and is $\frac{4}{51}$. This is an example of **conditional probability**.

PROBLEM What is the probability of picking an 8 from a well-shuffled deck of cards on the second draw, if the first card selected (and not returned to the deck) is an 8?

SOLUTION $P(8|8) = \frac{3}{51} = \frac{1}{17}$

PROBLEM What is the probability of drawing a king first and an 8 second from two draws of the deck, if the first card drawn is not replaced?

SOLUTION $P(K \text{ and } 8) = P(K) \times P(8|K) = \left(\frac{1}{13}\right)\left(\frac{4}{51}\right) = \frac{4}{663}$

Two events, A and B, are said to be **independent** of each other, if $P(B|A) = P(B)$. That is, the first outcome has no impact on the opportunity for the second outcome to occur. A consequence of this is that two events are independent if and only if $P(A \text{ and } B) = P(A) \times P(B)$.

PROBLEM A fair six-sided die (with all sides having the same chance of landing facing up) is rolled twice. What is the probability that the second roll is a six, given that the first roll is a five?

SOLUTION Unlike pulling a card from a deck and not returning the card to the deck, the chance of rolling a six is the same whether it is the first roll or the tenth roll.

$$P(6|5) = \frac{1}{6}$$

PROBLEM The table shows the result of a random sample of 200 voters. The voters were asked to identify how they were registered with the local Board of Elections.

GENDER	DEMOCRAT	REPUBLICAN	INDEPENDENT	TOTAL
Female	31	29	64	124
Male	19	31	26	76
Total	50	60	90	200

a. If a voter from this survey is selected at random, what is the probability that the voter is a registered Democrat?

b. If a voter from this survey is selected at random, what is the probability that the voter is female?

c. If a voter from this survey is selected at random, what is the probability that the voter is a registered Democrat given that the voter is a woman?

d. If a voter from this survey is selected at random, what is the probability that the voter is a woman given that the voter is a registered Democrat?

SOLUTION a. Fifty of the people in the survey are registered Democrats, so $P(\text{Democrat}) = \frac{50}{200} = \frac{1}{4}$.

b. Of the people who took the survey, 124 of them are women, so $P(\text{Female}) = \frac{124}{200} = \frac{31}{50}$.

c. Of the 124 women who took the survey, 31 are Democrats.

$P(\text{Democrat}|\text{Female}) = \frac{31}{124} = \frac{1}{4}$.

d. Of the 50 Democrats, 31 are women. $P(\text{Female}\,|\,\text{Democrat}) = \frac{31}{50}$.

The two events "the voter is a woman" and "the voter is a registered Democrat" are independent of each other.

A second test for independence is the rule: if $P(A \text{ and } B) = P(A) \times P(B)$, then A and B are independent.

PROBLEM The results of a survey of the junior class at Central High School are shown in the Venn diagram.

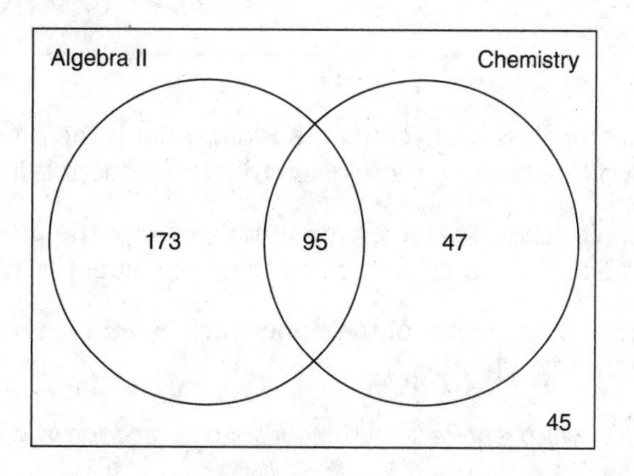

a. How many students are in the junior class of Central High School?

b. If a junior from Central High School is selected at random, what is the probability the student is enrolled in an Algebra II class?

c. If a junior from Central High school is selected at random, what is the probability the student is enrolled in an Algebra II class given that the student is enrolled in a chemistry class?

d. Are the events "the student is enrolled in Algebra II" and "the student is enrolled in chemistry" independent events for the juniors at Central High School?

SOLUTION a. There are $173 + 95 + 47 + 45 = 360$ juniors in Central High School.

b. $P(\text{Algebra II}) = \frac{173+95}{360} = \frac{268}{360}$

c. $P(\text{Algebra II}\,|\,\text{chemistry}) = \frac{95}{47+95} = \frac{95}{142}$

d. Because $P(\text{Algebra II}\,|\,\text{chemistry}) \neq P(\text{Algebra II})$, the two events are not independent.

Solve the following.

1. What is the probability that the second card drawn from a standard bridge deck of cards is a spade given that the first card drawn (and not returned to the deck) is a heart?

Use the accompanying Venn diagram, which represents the members of the senior class at a small high school, to answer questions 2–4.

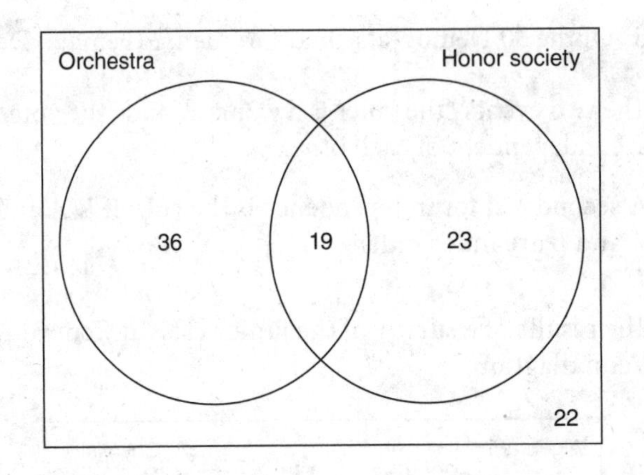

2. If a member of the senior class is selected at random, what is the probability that the student is a member of the honor society given that the student is in orchestra?
3. If a member of the senior class is selected at random, what is the probability that the student is not a member of the orchestra given that the student is in the honor society?
4. Are the events "a senior is a member of the honor society" and "a senior is a member of the orchestra" independent of each other?

Use the accompanying table, which represents the amount of time a person listens to music each day, to answer questions 5–7.

AGE/HOURS	1-2 HR	2-3 HR	4+ HR	TOTAL
15–17	20	23	29	72
18–20	19	22	32	73
21–24	17	34	27	78
Total	56	79	88	223

5. What is the probability that a randomly selected person from this survey listens to music 1–2 hr each day, if that person is between the ages of 15 and 17?
6. What is the probability that a randomly selected person from this survey is between the ages of 18 and 20, if that person listens to music for 4 or more hours each day?
7. Are the events "the person is between 21 and 24 years old" and "the person listens to music 2–3 hr each day" independent of each other?

Binomial probability/Bernoulli trial

A probability experiment with the properties that

- there are n independent trials;
- there are two outcomes per trial (success and failure); and,
- P(success) is constant from trial to trial

is called a **binomial experiment**. (It is also called a **Bernoulli trial** after the family of Swiss mathematicians.) The probability of exactly r successes in n trials is given by the formula $\binom{n}{r}p^r(1-p)^{n-r}$, where p represents the probability of success on any given trial.

PROBLEM A fair six-sided die is rolled five times. What is the probability of getting a one three times?

SOLUTION If success is defined as "the die shows a one," and failure by "the die does not show a one," then this is a binomial experiment.

This is key—it is not important if the roll is a two or five, whether or not it is a one is the only thing that matters.

Since this is a binomial experiment, $P(r=3)=\binom{5}{3}\left(\frac{1}{6}\right)^3\left(\frac{5}{6}\right)^2=\frac{250}{7776}$.

There is a function on most graphing calculators called binomPdf that will compute the probability of a specific number of events. Look at the directions for your graphing device to see if you have it.

PROBLEM Three percent of all grommets produced by the Ace Corporation are defective. In a random sample of 100 grommets produced by Ace, what is the probability that at most three of them are defective?

SOLUTION This is also a binomial experiment, because the chance of a grommet being defective is independent of whether the previous grommet tested was defective. Success in this experiment is defined by finding a defective grommet. With $n=100$ and $p=0.05$:

$$P(r\le 3)=P(r=0)+P(r=1)+P(r=2)+P(r=3)$$

$$=\binom{100}{0}(0.03)^0(0.97)^{100}+\binom{100}{1}(0.03)^1(0.97)^{99}+\binom{100}{2}(0.03)^2(0.97)^{98}$$

$$+\binom{100}{3}(0.03)^3(0.97)^{97}=0.6472\text{, rounded to four decimal places.}$$

There is a function on most graphing calculators called binomCdf that will compute the probability of a range of events. Look at the directions for your graphing device to see if you have it.

PROBLEM The spinner shown in the accompanying diagram is spun 10 times. What is the probability of having the outcome be green at least three times?

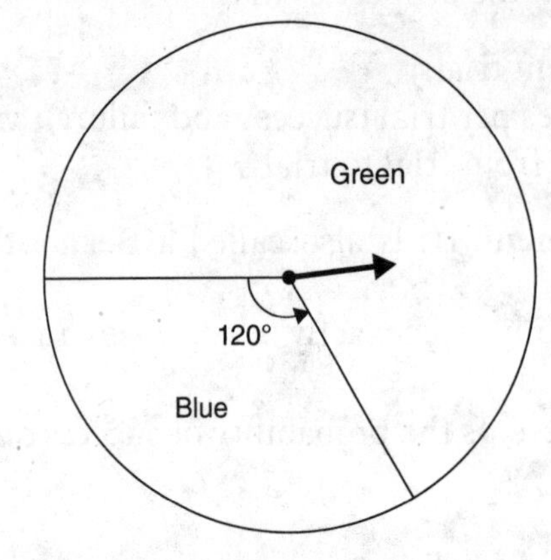

SOLUTION The board does not change shape, so $P(\text{green})$ is consistently $\frac{2}{3}$.

$P(r \geq 3) = P(r = 3) + P(r = 4) + P(r = 5) + \cdots + P(r = 10)$

This is a lot of computing to do. The complement of this event—the outcomes that are not the event—are $P(r \leq 2) = P(r = 0) + P(r = 1) + P(r = 2)$. Take advantage of the fact that $P(\text{event}) + P(\text{event's complement})$ must equal 1 because, together, the event and its complement make up all the possible outcomes for an experiment.

$P(r \geq 3) = 1 - P(r \leq 2) =$

$$1 - \left(\binom{10}{0}\left(\frac{2}{3}\right)^0\left(\frac{1}{3}\right)^{10} + \binom{10}{1}\left(\frac{2}{3}\right)^1\left(\frac{1}{3}\right)^9 + \binom{10}{2}\left(\frac{2}{3}\right)^2\left(\frac{1}{3}\right)^8 \right) = 0.7009,$$

answer rounded to four decimal places.

EXERCISE

9·6

Use the diagram of the spinner with six congruent regions to answer questions 1–3.

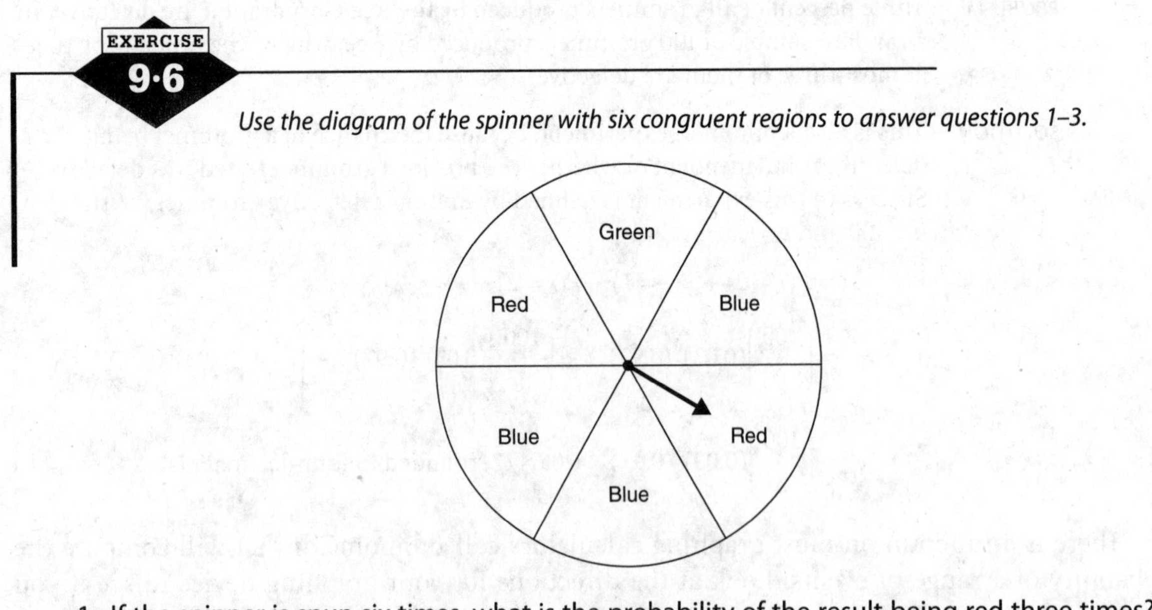

1. If the spinner is spun six times, what is the probability of the result being red three times?
2. If the spinner is spun five times, what is the probability of the result being blue at most twice?
3. If the spinner is spun 10 times, what is the probability of the result being green at least twice?

4. Based on past statistics, a company knows that 99% of its ball bearings pass a quality control test. A random sample of 100 ball bearings is tested for the quality control test.
   (a) What is the probability that all 100 bearings will pass the test?
   (b) What is the probability that exactly 2 of the 100 bearings will *fail* the test?
5. Laura is one of the school's best tennis players. Laura's rate of winning first-serve points is 78%.
   (a) What is the probability that Laura will win half of her first-serve points in her next 10 serves?
   (b) What is the probability that Laura will win at most 8 first-serve points in her next 10 serves?
   (c) What is the probability that Laura will win at least 9 of her first-serve points in her next 10 serves?

Introduction to statistics

Descriptive statistics gives information about the center, spread, and shape of a set of data. The relationship between center and spread is very important. Consider two students who each score an 85 on a test. If the range of grades for the test taken by the first student is 82 to 89 while the range of grades for the test taken by the second student is 48 to 87, who did better?

Inferential statistics is the science of making statements about a larger group (the population) from information gathered from a smaller group (the sample). In this chapter, you will look at the normal distribution, a key piece in the study of inferential statistics, and regression, to find equations that relate bivariate data.

Measures of central tendency

The two most popular measures of central tendency are the **mean** and the **median**. The mean of a data set is found by adding all numbers in the data set and then dividing by the number of values in the set. The formula is $\frac{1}{n}\sum_{i=1}^{n} X_i$, where X_i represents each element in the data set. If the data represent the entire population, the result is called a **parameter**. The population mean is represented by the Greek letter mu, μ. If the data represent a sample of the population, the result is called a **statistic**. The sample mean is represented by the variable of the data with a bar above it. In this case, the mean is represented by $\overline{x}$.

If the data are arranged in increasing or decreasing order, the number in the middle is called the median. If there is an odd number of data points, such as 3, 6, 7, 9, 11, then there will be the same number of data points below the center point (3, 6) as there are above it (9, 11). In this case, the middle point in the range of data, or the median, is 7. If there is an even number of data points, such as 3, 6, 7, 9, 11, 14, there are the same number of data points in the "lower" grouping (3, 6, 7) as in the "upper" grouping (9, 11, 14). The median of this group is equal to the mean of the two middle numbers, $\left(\frac{7+9}{2}\right) = 8$. Please observe that the median is not one of the original data points.

While the common usage of the word *average* implies the mean, average, as a measure of center, could also be the median.

PROBLEM The following data represent the total salaries of the 32 NFL teams for the 2007 season (totals rounded to the nearest tenth of a million dollars). Determine the mean and median of the data.

TEAM	TOTAL PAYROLL	TEAM	TOTAL PAYROLL
Washington Redskins	$123.4	St. Louis Rams	$100.3
New England Patriots	$118.0	New York Jets	$100.0
New Orleans Saints	$110.4	Seattle Seahawks	$99.6
Buffalo Bills	$108.9	Arizona Cardinals	$98.7
Kansas City Chiefs	$108.5	Cincinnati Bengals	$98.5
Dallas Cowboys	$107.4	Houston Texans	$98.2
San Francisco 49ers	$106.9	Tampa Bay Buccaneers	$98.1
Detroit Lions	$106.7	Green Bay Packers	$97.7
Pittsburgh Steelers	$106.3	Tennessee Titans	$97.1
Baltimore Ravens	$105.0	Jacksonville Jaguars	$94.0
Chicago Bears	$104.2	Carolina Panthers	$93.9
Indianapolis Colts	$102.8	Miami Dolphins	$92.6
San Diego Chargers	$102.5	Minnesota Vikings	$92.2
Cleveland Browns	$102.4	Oakland Raiders	$90.9
Denver Broncos	$102.2	Atlanta Falcons	$83.8
Philadelphia Eagles	$100.8	New York Giants	$75.8

Data from http://content.usatoday.com/sportsdata/football/nfl/salaries/team/2007

SOLUTION The data are presented in decreasing order of salaries. The median for these data will be the mean of the sixteenth and seventeenth ranked teams (Philadelphia and St. Louis). The median is $100.55 million. The mean for the data can best be computed with technology. Enter the data into a list on your graphing calculator or the column of a spreadsheet. Because these data represent the entire population of teams in the NFL, the population mean is $\mu = \$100.87$ million.

It should be mentioned here that the median is less impacted by an unusually large or small number than is the mean. Suppose that the total team salary for the Washington Redskins was $1023.4 million rather than $123.4 million. The median for the salaries would remain at $100.55 million while the mean for the data would increase to $128.99 million.

PROBLEM A restaurant manager kept tabs on the number of dinners served each night for a total of 854 nights. The data are presented in the table. Determine the median and mean for these data.

DINNERS	COUNT
33	30
34	38
35	51
38	85
40	120
45	150
48	170
50	210

SOLUTION There are 854 pieces of data. Divide this number by 2 to get 427. The median of the data is the mean of the 427th and 428th pieces of data. Sum the count values to discover in which row of the table these pieces of data would be found. The sum of the first five numbers in the count column is 324($30+38+51+85+120=324$), so only the first 324 data points are represented by these five values. Adding the next two values ($324+150$) results in a data count of 474. Therefore, the 427th and 428th pieces of data are both 45, so the median of the data set is 45 dinners per night.

To compute the mean, the sum of the dinners served on each of the 854 nights must be computed. That is, $\bar{x}=\dfrac{(33)(30)+(34)(38)+\cdots+(50)(210)}{854}$. Use your graphing calculator or computer program to do this computation. With the TI 83/84 and Nspire series, this would be a one variable statistical calculation with Dinners as the data values and Count as the frequency. The mean for this sample data is $\bar{x}=43.9$ dinners per night.

EXERCISE 10·1

Determine the mean and median for each of the following questions, using the data provided.

1. A random sample of measurements (in mm) of the diameter of 20 ball bearings is recorded as being:

   14.9, 11.4, 11.4, 10.6, 10.2, 13.6, 10.0, 12.1, 11.5, 14.9,
   10.1, 14.2, 13.1, 11.0, 15.0, 13.6, 11.5, 11.2, 14.8, 14.7

2. Ten measurements are made for the time (to the nearest hundredth of a second) of the period of a pendulum. The measurements are:

   10.11, 10.24, 10.86, 10.37, 10.6, 10.61, 10.07, 10.01, 10.01, 10.7

3. The grades of 175 students on a state-wide mathematics test are displayed in the table.

GRADE	FREQUENCY	GRADE	FREQUENCY
50	3	80	30
60	9	85	27
65	12	90	28
70	20	95	19
75	23	100	4

4. A light bulb manufacturer tests a random selection of 457 light bulbs to determine the average life span of the bulb. The amount of time each bulb stayed lit (rounded to the nearest 25 hours) is given in the table.

HOURS	FREQUENCY	HOURS	FREQUENCY
850	10	1000	70
875	14	1025	60
900	23	1050	65
925	40	1075	40
950	50	1100	30
975	55		

Measures of dispersion

As was noted in the section on the measures of central tendency, an unusually large or small data value will impact the mean but not the median of a set of data. To help clarify the nature of the data being summarized, the mean is usually reported with a second number representing how the data are dispersed (spread out). The simplest measure of dispersion is the range of the data. The range is the difference between the maximum and minimum values in the data set. Two more useful measures of dispersion (useful, in that they are used in the branch of statistics called inferential statistics) are the interquartile range and the standard deviation.

The **Interquartile Range (IQR)** represents the middle 50% of the values for a set of data. The **first quartile (Q1)** is the 25th percentile and the **third quartile (Q3)** is the 75th percentile. The IQR is the difference of Q3 and Q1. (Q1 is midway between the minimum value and the median, while Q3 is midway between the median and the maximum value.) The IQR is used to determine whether a data value is "unusually" large or small. The rule of thumb is that a data value that is more than 1.5 times the IQR greater than Q3 is considered unusually large, while a number that is less than 1.5 times the IQR below Q1 is considered unusually small. Numbers that fit this definition of unusually large or small are called **outliers**.

The **box-and-whisker plot** is a graphical representation of data which displays the minimum, Q1, median, Q3, and maximum values of the data set. The five numbers are referred to as the **five-number summary** for the data.

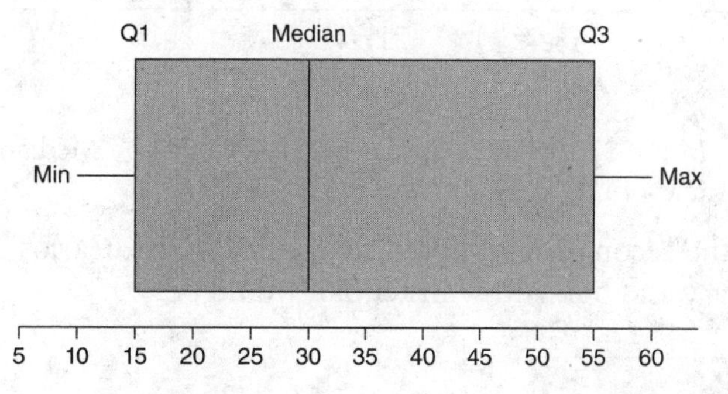

PROBLEM The following table contains the names and salaries for 23 of the top 25 players in major league baseball for the 2002 season. Compute the five-number summary and construct the box-and-whisker plot for the 23 known players. (The salary data for Carlos Delgado and Alex Rodriguez will be discussed following the solution to this problem.)

PLAYER	SALARY	PLAYER	SALARY
Rodriguez, Alex	$$$$$$$	Williams, Bernie	$12,357,143
Delgado, Carlos	$$$$$$	Vaughn, Mo	$12,166,667
Brown, Kevin	$15,714,286	Jones, Chipper	$11,333,333
Ramirez, Manny	$15,462,727	Bagwell, Jeff	$11,000,000
Bonds, Barry	$15,000,000	Gonzalez, Juan	$11,000,000
Sosa, Sammy	$15,000,000	Mondesi, Raul	$11,000,000
Jeter, Derek	$14,600,000	Mussina, Mike	$11,000,000
Martinez, Pedro	$14,000,000	Piazza, Mike	$10,571,429
Green, Shawn	$13,416,667	Giambi, Jason	$10,428,571
Johnson, Randy	$13,350,000	Clemens, Roger	$10,300,000
Maddux, Greg	$13,100,000	Jones, Andruw	$10,000,000

(continued)

PLAYER	SALARY	PLAYER	SALARY
Walker, Larry	$12,666,667	Schilling, Curt	$10,000,000
Belle, Albert	$12,368,790		

Data from http://content.usatoday.com/sportsdata/baseball/mlb/salaries/player/top-25/2002 box-and-whisker plot

SOLUTION

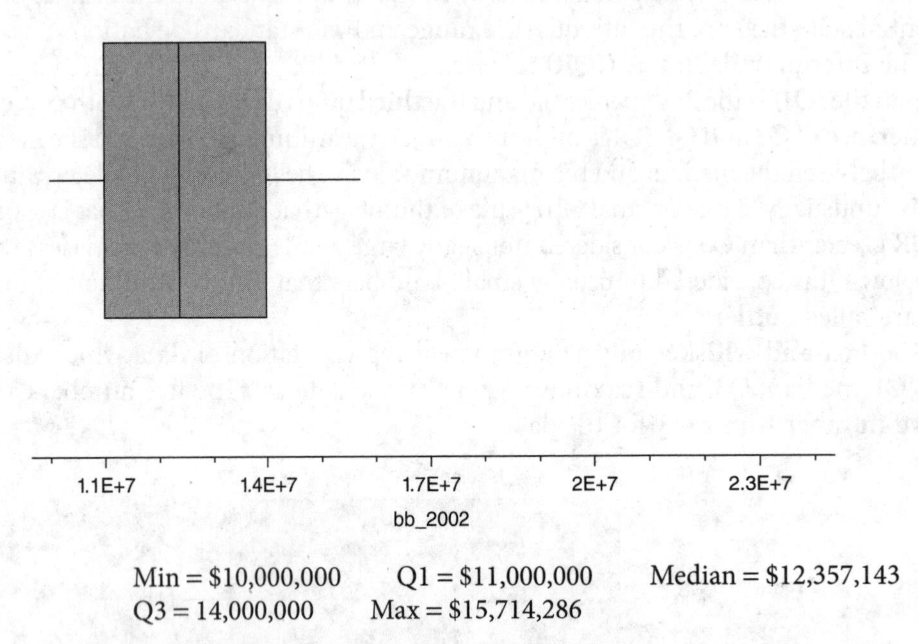

Min = $10,000,000 Q1 = $11,000,000 Median = $12,357,143
Q3 = 14,000,000 Max = $15,714,286

If we include the second-highest-paid salary in the list, Carlos Delgado's $19,400,000, the five-number summary and box-and-whisker plot would be:

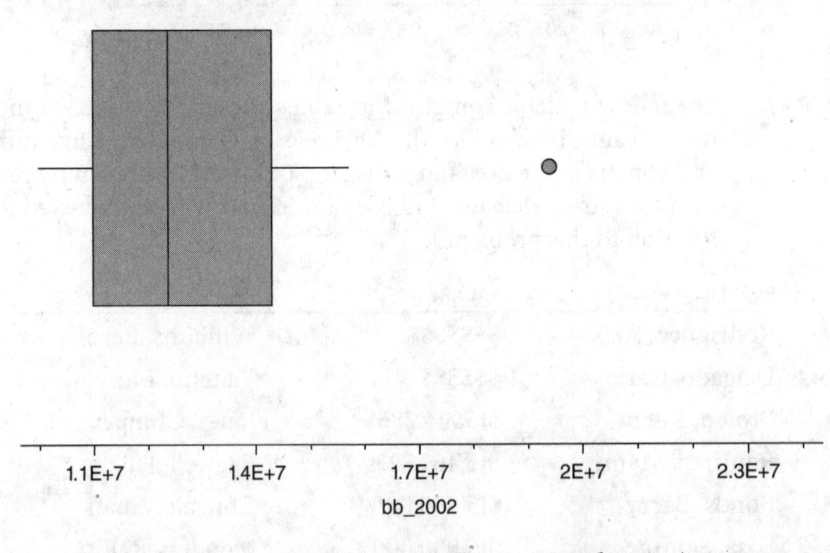

Min = $10,000,000 Q1 = $11,000,000 Median = $12,362,966.50
Q3 = 14,300,000 Max = $19,400,000

With this change in the maximum value considered, there is a slight difference in the values of the median and third quartile. Also note that Delgado's salary is so much larger than the previous maximum that it is not connected to the box-and-whisker plot. Delgado's salary is an outlier for the data. For these data, the IQR is $3,300,000. Multiplied by 1.5, this becomes $4,950,000. Delgado's salary is $5,100,000 more than the Q3 value of $14,300,000, so it is considered an outlier.

The number one salary for that year, $22,000,000, was earned by Alex Rodriguez. The impact this number has on the five-number summary and on the box-and-whisker plot is interesting. Delgado's salary is no longer an outlier.

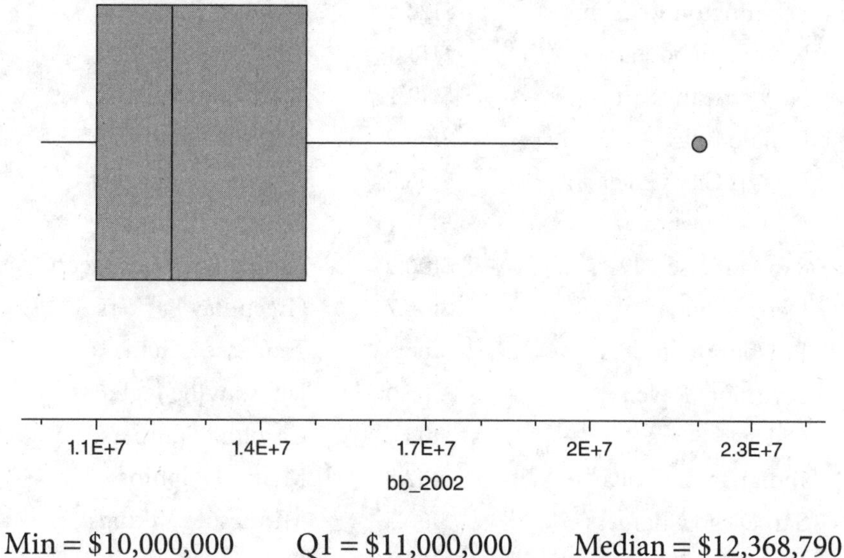

Min = $10,000,000 Q1 = $11,000,000 Median = $12,368,790
Q3 = 14,800,000 Max = $22,000,000

The other important measure of dispersion is the **standard deviation**. This measures, on average, how far each data point is from the mean of the data set. We need to be careful because this measure can be used with a sample to make inferences about a population. There is a difference between the method for computing the standard deviation for a sample and the standard deviation for a population. The sample standard deviation, designated by s, is computed by adding the squares of the differences between each data value and the mean, divide by one less than the number of pieces of data in the set, and then take the square root of the quotient. Written as a formula, we read $s = \sqrt{\dfrac{\sum_{i=1}^{n}(x_i - \bar{x})^2}{n-1}}$. The population standard deviation, designated by σ, is almost the same number. The sum of the squares is divided by the number of pieces of data in the data set, rather than by one less than this number. That is, $\sigma = \sqrt{\dfrac{\sum_{i=1}^{n}(x_i - \bar{x})^2}{n}}$.

Why is there a difference between the formula for the sample standard deviation and the formula for the population standard deviation? A more in-depth study of statistics shows that one gets a better prediction of the population from a sample when the sample standard deviation formula with the divisor being $n-1$ is used than when the divisor is n. If this has piqued your curiosity, perform an Internet search on degrees of freedom and you may gain a little more insight into an important theory of statistics.

So far, we have two statistical measures for samples and populations that go by the same name and have different symbols.

SAMPLE STATISTIC		POPULATION PARAMETER
$\bar{x}$	Mean	μ
s	Standard deviation	σ

PROBLEM Compute the standard deviation for the salaries of the 32 NFL teams during the 2007 season.

TEAM	TOTAL PAYROLL	TEAM	TOTAL PAYROLL
Washington Redskins	$123.4	St. Louis Rams	$100.3
New England Patriots	$118.0	New York Jets	$100.0
New Orleans Saints	$110.4	Seattle Seahawks	$99.6
Buffalo Bills	$108.9	Arizona Cardinals	$98.7
Kansas City Chiefs	$108.5	Cincinnati Bengals	$98.5
Dallas Cowboys	$107.4	Houston Texans	$98.2
San Francisco 49ers	$106.9	Tampa Bay Buccaneers	$98.1
Detroit Lions	$106.7	Green Bay Packers	$97.7
Pittsburgh Steelers	$106.3	Tennessee Titans	$97.1
Baltimore Ravens	$105.0	Jacksonville Jaguars	$94.0
Chicago Bears	$104.2	Carolina Panthers	$93.9
Indianapolis Colts	$102.8	Miami Dolphins	$92.6
San Diego Chargers	$102.5	Minnesota Vikings	$92.2
Cleveland Browns	$102.4	Oakland Raiders	$90.9
Denver Broncos	$102.2	Atlanta Falcons	$83.8
Philadelphia Eagles	$100.8	New York Giants	$75.8

Data from http://content.usatoday.com/sportsdata/football/nfl/salaries/team/2007

SOLUTION Enter the data into your graphing calculator or computing device. Because these data represent the entire population for the NFL, compute σ to equal $8.85 million.

PROBLEM A restaurant manager kept tabs on the number of dinners served each night for a total of 854 nights. The data are presented in the following table. Compute the standard deviation for the number of dinners served at the restaurant during the 854 days.

DINNERS	COUNT
33	30
34	38
35	51
38	85
40	120
45	150
48	170
50	210

SOLUTION Enter the data into your calculator or computing device and compute the sample standard deviation, *s*. (The data represent a sample of the number of dinners served in the restaurant.) Your result should be 5.6 dinners. Be sure to remember to identify the frequency of the data as the Count value, not as 1.

Given the data provided for each question, compute the values specified.

1. The Pacific salmon (*Oncorhynchus* spp.) spends part of its life in freshwater and the other in saltwater. A fisheries scientist at the University of Washington was interested in the distance that salmon travel per day while at sea. She radio-tagged 16 randomly selected salmon and determined the miles traveled per day. Her findings are given in the following table. Determine the interquartile range and standard deviation for these data.

31	16	29	13	21	23	38	30
44	51	38	22	35	30	18	28

Data from Public Broadcasting System

2. The top 20 average home game attendance figures for the National Football League in 2010 are listed in the following table.

TEAM	AVERAGE ATTENDANCE	TEAM	AVERAGE ATTENDANCE
Dallas	87,047	San Francisco	69,732
Washington	83,172	Philadelphia	69,144
New York Giants	79,019	Tennessee	69,143
New York Jets	78,596	New England	68,756
Denver	74,908	Atlanta	67,850
Carolina	72,620	Miami	67,744
Baltimore	71,727	Kansas City	67,672
Houston	71,080	Seattle	66,992
Green Bay	70,795	Indianapolis	66,975
New Orleans	70,038	Cleveland	66,116

Data from espn.com.

a. Compute the mean and median for these data.
b. Compute the interquartile range and the standard deviation for these data.

3. The following table displays the number of calories in all McDonald's sandwiches.

SANDWICH	CALORIES
Hamburger 3.5 oz (100g)	250
Cheeseburger 4 oz (114 g)	300
Double Cheeseburger 5.8 oz (165 g)	440
McDouble 5.3 oz (151 g)	390
Quarter Pounder® with Cheese 7 oz (198 g)	510
Double Quarter Pounder® with Cheese 9.8 oz (279 g)	740
Big Mac® 7.5 oz (214 g)	540
Big N' Tasty® 7.2 oz (206 g)	460
Big N' Tasty® with Cheese 7.7 oz (220 g)	510
Angus Bacon & Cheese 10.2 oz (291 g)	790
Angus Deluxe 11.1 oz (314 g)	750

(continued)

SANDWICH	CALORIES
Angus Mushroom & Swiss 10 oz (283 g)	770
Filet-O-Fish® 5 oz (142 g)	380
McChicken® 5 oz (143 g)	360
McRib® 7.4 oz (209 g)	500
Premium Grilled Chicken Classic Sandwich 7 oz (200 g)	350
Premium Crispy Chicken Classic Sandwich 7.5 oz (213 g)	510
Premium Grilled Chicken Club Sandwich 7.9 oz (223 g)	460
Premium Crispy Chicken Club Sandwich 8.4 oz (237 g)	620
Premium Grilled Chicken Ranch BLT Sandwich 7.1 oz (202 g)	380
Premium Crispy Chicken Ranch BLT Sandwich 7.6 oz (215 g)	540
Southern Style Crispy Chicken Sandwich 8.3 oz (155 g)	420
Ranch Snack Wrap® (Crispy) 4.2 oz (118 g)	350
Ranch Snack Wrap® (Grilled) 4.2 oz (118 g)	270
Honey Mustard Snack Wrap® (Crispy) 4.1 oz (116 g)	330
Honey Mustard Snack Wrap® (Grilled) 4.1 oz (116 g)	250
Chipotle BBQ Snack Wrap® (Crispy) 4.1 oz (117 g)	330
Chipotle BBQ Snack Wrap® (Grilled) 4.2 oz (117 g)	250
Angus Bacon & Cheese Snack Wrap 5.1 oz (145 g)	390
Angus Chipotle BBQ Bacon 10.3 oz (294 g)	800
Angus Chipotle BBQ Bacon Snack Wrap 5.2 oz (148 g)	400
Angus Deluxe Snack Wrap 6 oz (170 g)	410
Angus Mushroom & Swiss Snack Wrap 5.7 oz (162 g)	430
Mac Snack Wrap 4.4 oz (126 g)	330

Data from http://nutrition.mcdonalds.com/nutritionexchange/nutritionfacts.pdf

a. Compute the mean and median for these data.
b. Compute the interquartile range and the standard deviation for these data.

4. The following information was gathered about the number of calories from the 15 sandwiches available from Burger King (data from http://www.fastfoodnutrition.org/r-nutrition-facts/Burger%20King-item.html).

   Mean: 616 calories; Median: 640 calories; Min.: 340 calories; Q1: 430 calories; Q3: 710 calories; Max.: 1020 calories; Population Standard Deviation: 187.9 calories

   Compare these data against the data from the McDonald's sandwiches with regard to the measures of center and spread.

Normal distribution

The probability for a **continuous random variable** (a variable which is measured rather than counted) is computed by measuring the area under a density curve. When you study calculus, you will learn other strategies for finding the area under curves; for now we will use technology to determine this. Before we continue with the use of technology, there is a simple problem that you need to understand. The area under the bell curve gives the probability of an event. This means that two points—a left endpoint and a right endpoint—need to be located so that the area between these points can be determined. What if the left endpoint and the right endpoint are the same number? The result will be a line segment from the x-axis to the point on the graph of the bell curve. Since a line segment is a one-dimensional figure, it has no area. This illustrates an important aspect of the continuous random variable—probabilities are computed for an interval (however small the interval) but not for a point. That is, we can calculate the probability that x is between 1.5 and 2.5, between 1.9 and 2.1, between 1.999 and 2.001, but never at $x = 2$.

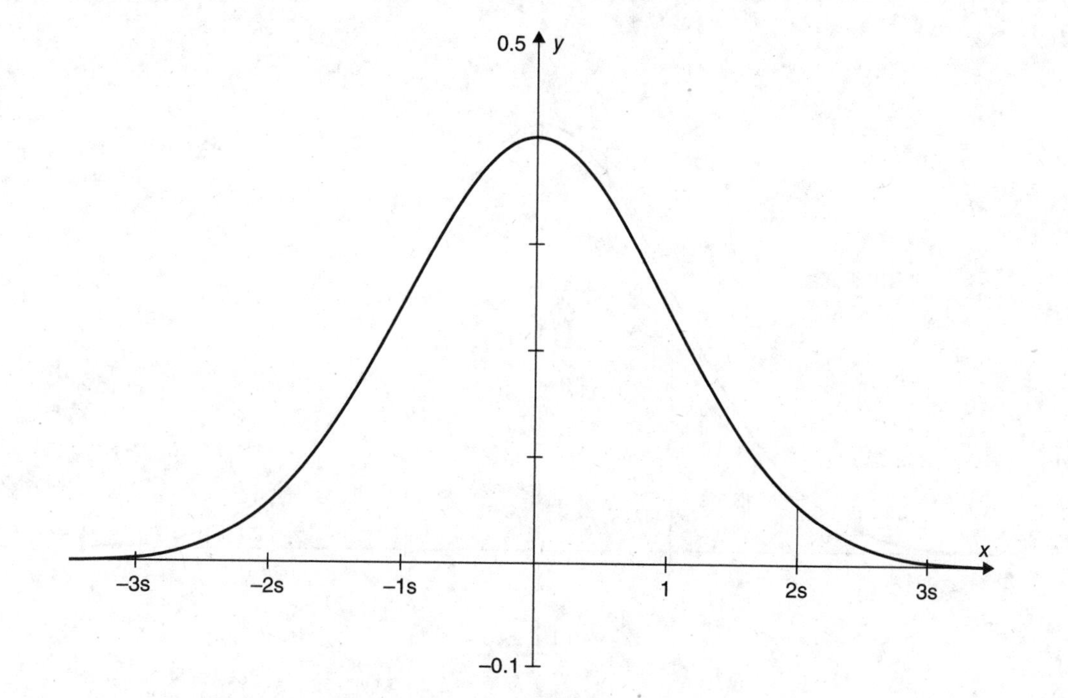

PROBLEM The height of students at Brookmount High School is normally distributed, with a mean of 67 inches, and a standard deviation of 3 inches. If a student is selected at random, what is the probability that the student's height will be between 61 and 73 inches?

PROBLEM The diameter of ball bearings is normally distributed, with a mean of 2.1 cm and a standard deviation of 0.15 cm. If a ball bearing is selected at random, what is the probability that its diameter will be between 1.8 and 2.4 cm?

SOLUTION Believe it or not, these two questions have the same answer. Reread the questions and observe that the lower bound for each region is two standard deviations less than the mean, while the upper bound is two standard deviations greater than the mean. Consequently, they will have the same area under the bell curve.

There are three benchmark probabilities for normal distributions that are considered to be common knowledge and are necessary for students of statistics.

- The first benchmark probability is that approximately 68% of the data lies within one standard deviation of the mean.
- The second benchmark probability is that approximately 95% of the data lies within two standard deviations of the mean.
- The third benchmark probability is that approximately 99.7% of the data lies within three standard deviations of the mean.

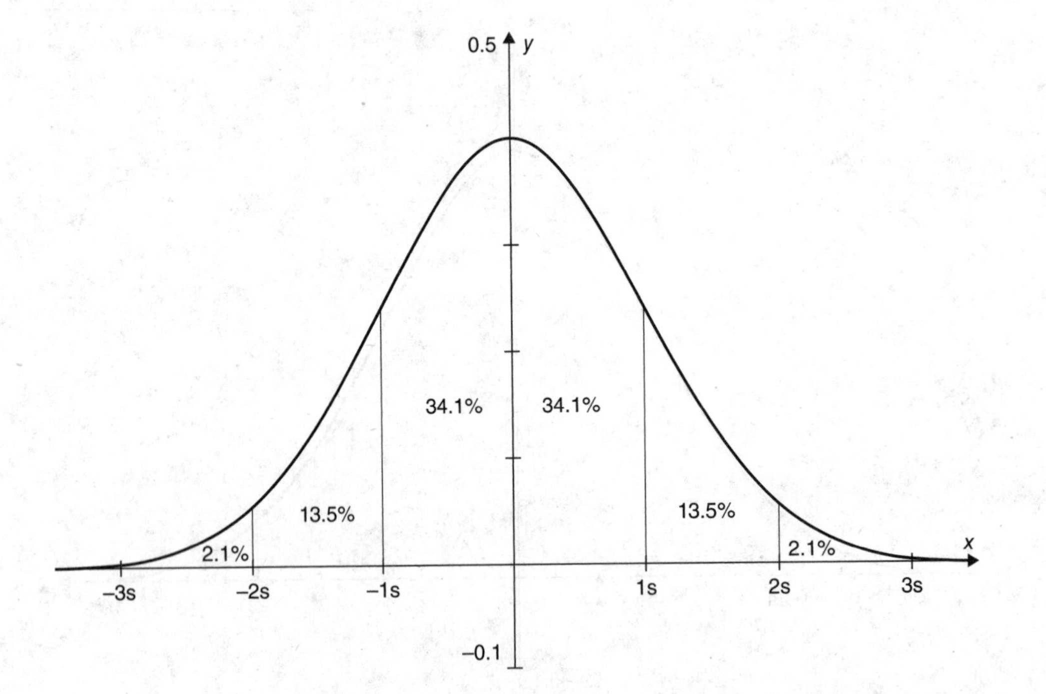

PROBLEM Let x be the variable of a normally distributed quantity with mean, μ, and standard deviation, σ. What is the probability that $x > \mu$ or, written symbolically, $P(x > \mu)$?

SOLUTION When you examine the area under the normal curve shown in the previous diagram, you see that it is symmetric about the mean, μ. Since the total area under the curve is 1 (because the area under the curve does represent all possibilities, the total area must be 1), half the area must be to the right of μ and half to the left of μ. Therefore, $P(x > \mu) = 0.5$.

PROBLEM The height of students at Brookmount High School is normally distributed with a mean of 67 inches and a standard deviation of 3 inches. If a student is selected at random, what is the probability that the student's height will be:

a. between 64 and 70 inches

b. between 67 and 73 inches

c. over 70 inches

d. under 73 inches

e. under 64 inches or over 70 inches

SOLUTION Let h represent the height of a randomly selected student.

a. $P(64 < h < 70) = 0.682$, because these numbers are each one standard deviation from the mean.

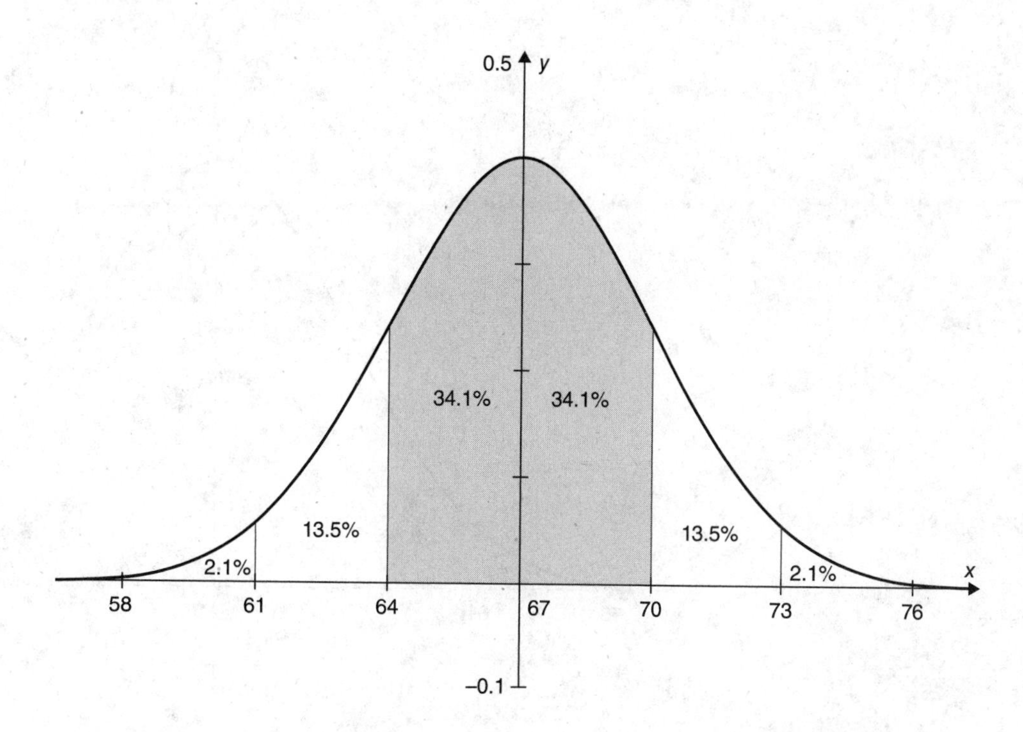

b. $P(67 < h < 73) = 0.476$ because this is the region from the mean to the point two standard deviations above the mean.

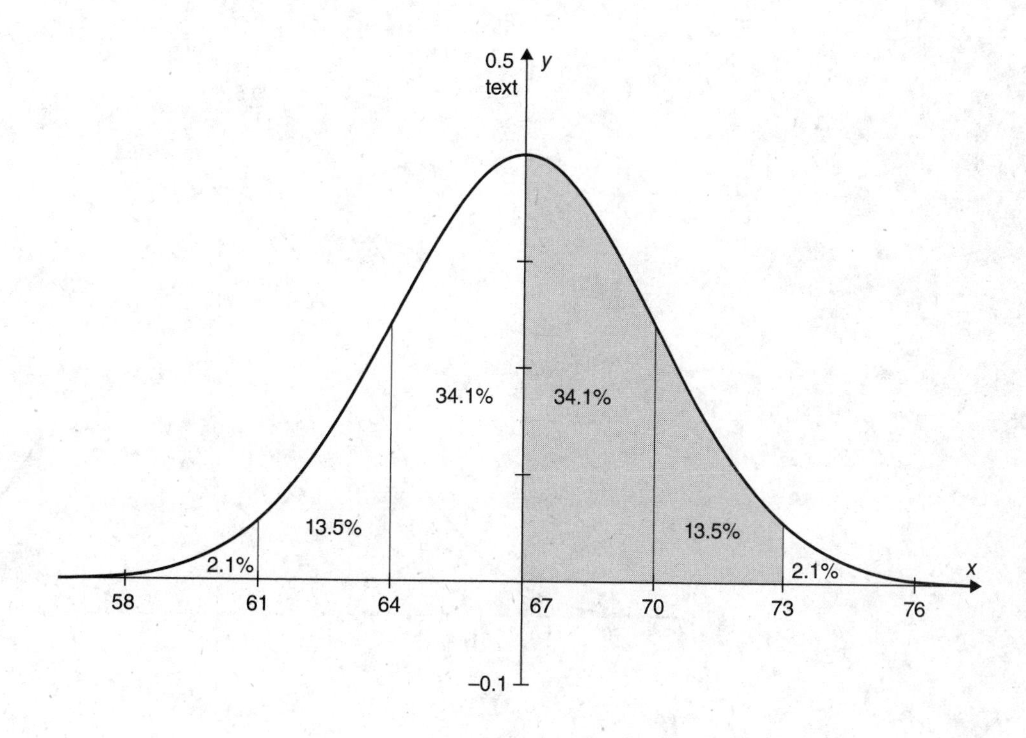

c. $P(h > 70)$ represents the region to the right of one standard deviation above the mean. Knowing that the area to the right of the mean is 0.5, and the area between the mean and one standard deviation above the mean is 0.341, $P(h > 70) = 0.5 - P(67 < h < 70) = 0.159$.

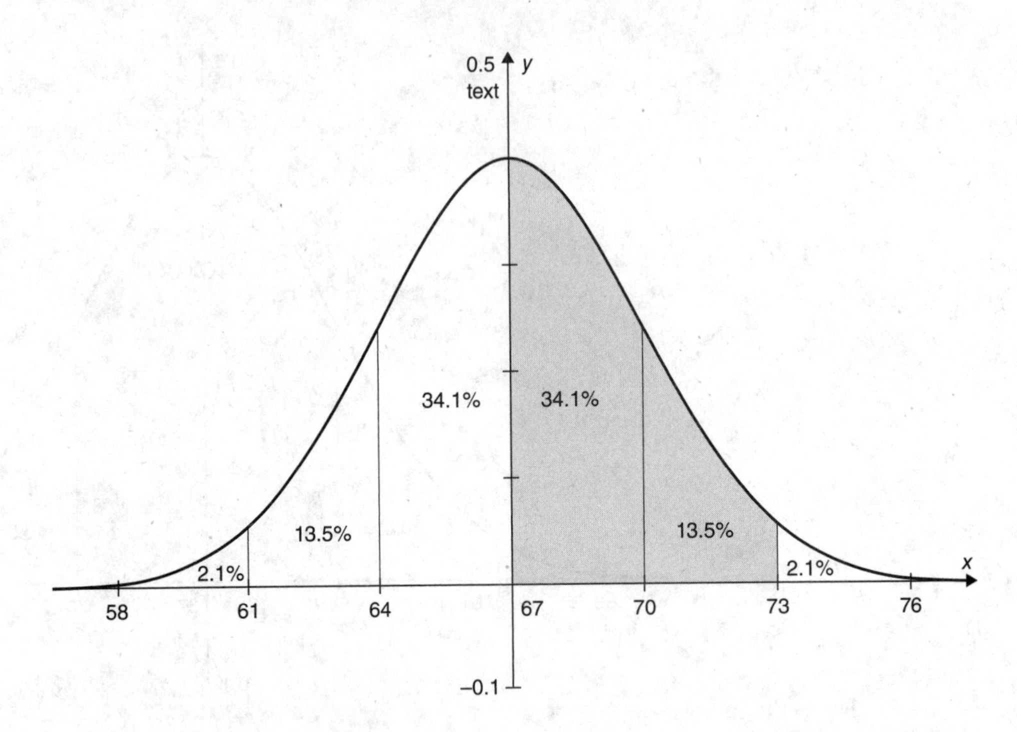

d. $P(h < 73)$ is the area to the left of 73. Since 73 represents the point two standard deviations above the mean, use $P(h < 73) = P(h < 67) + P(67 < h < 73) = 0.5 + 0.476 = 0.976$.

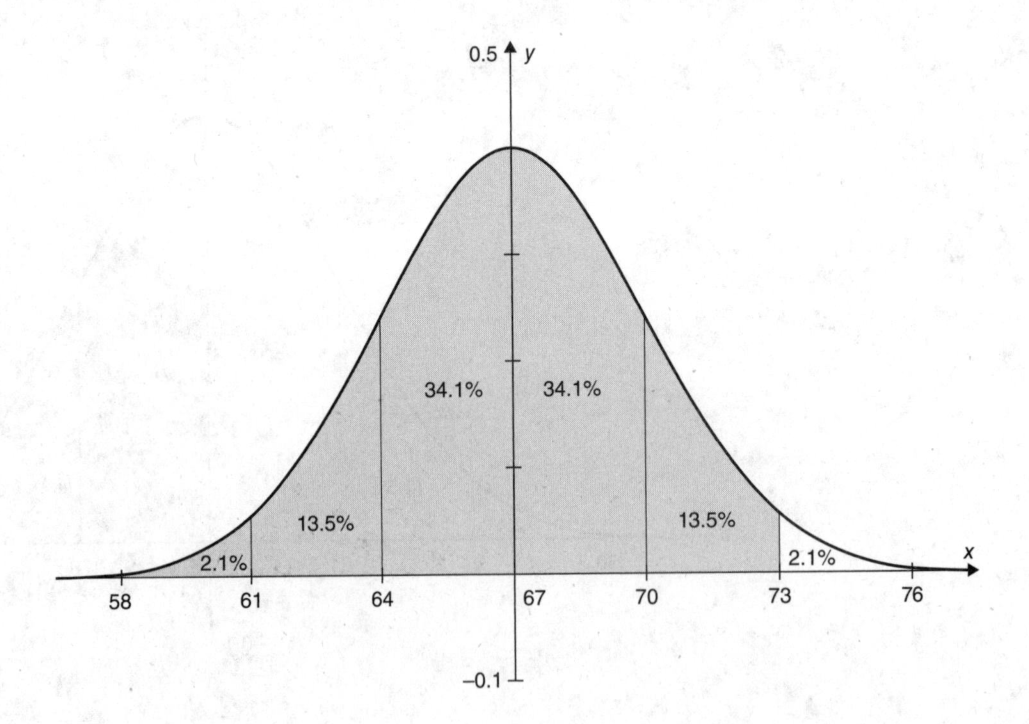

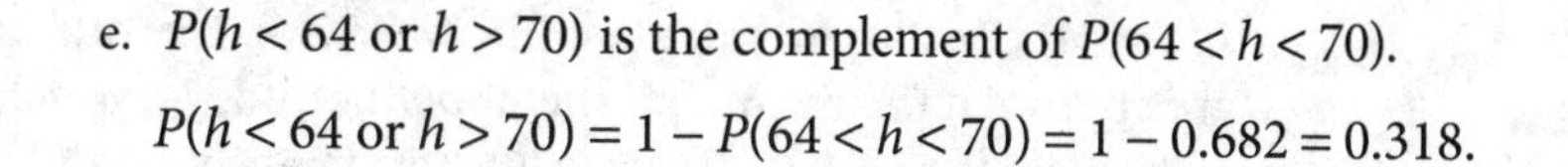

e. $P(h < 64 \text{ or } h > 70)$ is the complement of $P(64 < h < 70)$.

$P(h < 64 \text{ or } h > 70) = 1 - P(64 < h < 70) = 1 - 0.682 = 0.318$.

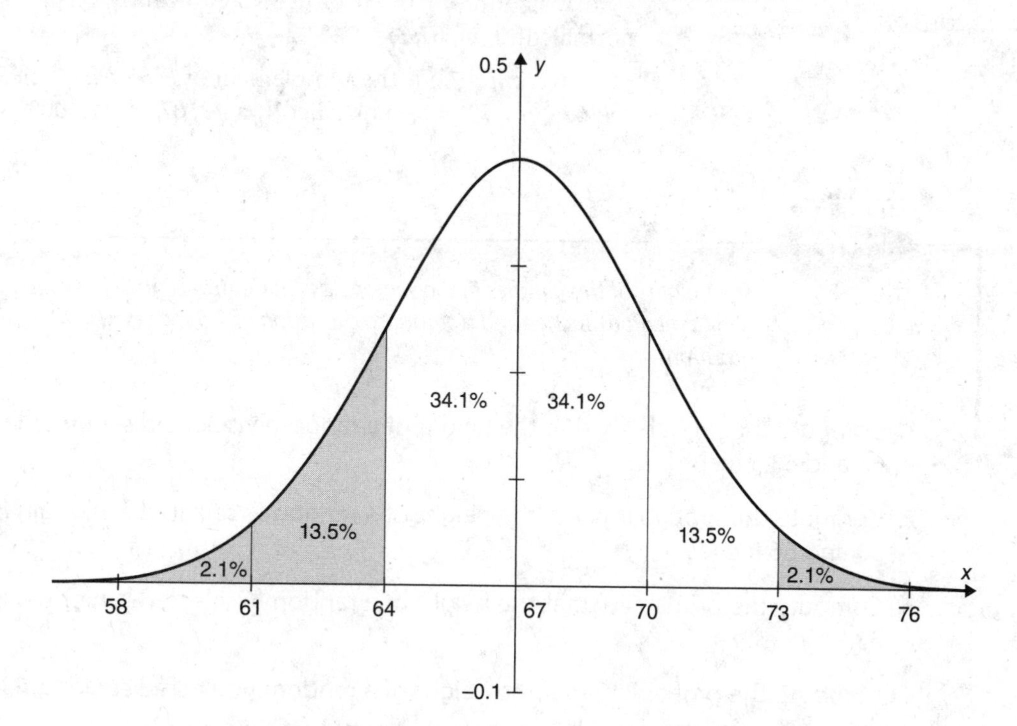

Most graphing calculators, spreadsheet programs, or computer programs with CAS capability have built-in functions that can compute probabilities. The TI 83/84 series and the Nspire calculators have a normcdf function that can compute probabilities for all values of the input variable. The parameters for this function are normcdf(lower bound, upper bound, mean, standard deviation).

PROBLEM Using a calculator or computer application, and the given distribution of heights of students at Brookmount High School, determine the probability that the height of a randomly selected student is:

a. between 66 and 71 inches

b. greater than 68.3 inches

c. less than 74 inches

d. less than 65 inches or greater than 72 inches

SOLUTION a. $P(66 < h < 71) = \text{normalcdf}(66, 71, 67, 3) = 0.5393$

b. $P(h > 68.3)$ can be calculated as one half minus the area between the mean and the lower bound, 68.3.

$0.5 - \text{normalcdf}(67, 68.3, 67, 3) = 0.3324$

$P(h > 68.3)$ can also be calculated with the technology by selecting an upper bound that is well beyond three deviations greater than the mean. For this example, three deviations larger than the mean, 67, is $67 + 3(3) = 76$. A number such as 85 is well above the three standard deviation value.

$\text{normalcdf}(68.3, 85, 67, 3) = 0.3324$

Try working with various upper bounds so that you get comfortable with this notion.

c. $P(h < 74)$ can also be calculated in two ways. Use one half plus the area between the mean and the upper bound, 74, which would be 0.5 + normalcdf(67, 74, 67, 3), or use a significantly small lower bound, such as normalcdf(0, 74, 67, 3) = 0.9902.

d. $P(h < 65 \text{ or } h > 72)$ is the complement of $P(65 < h < 72)$. $P(h < 65 \text{ or } h > 72) = 1 - P(65 < h < 72) = 1 -$ normalcdf(65, 72, 67, 3) = 0.3003.

EXERCISE

10·3

The heights of the females in the senior class at Central High School are normally distributed with a mean 61 inches and a standard deviation of 2.5 inches. Use these data to answer questions 1–6.

1. Compute the probability that the height of a randomly selected senior girl is between 61 and 63.5 inches.
2. Compute the probability that the height of a randomly selected senior girl is between 56 and 66 inches.
3. Compute the probability that the height of a randomly selected senior girl is greater than 63.5 inches.
4. Compute the probability that the height of a randomly selected senior girl is less than 66 inches.
5. Compute the probability that the height of a randomly selected senior girl is less than 56 inches or greater than 66 inches.
6. Compute the probability that the height of a randomly selected senior girl is less than 58.5 inches or greater than 66 inches.

A soda dispenser dispenses soda amounts with a normal distribution, a mean of 11.9 fluid ounces, and a standard deviation of 0.35 ounces. A cup of soda is randomly selected from all the cups of soda that have been dispensed. Use this information to answer questions 7–10.

7. What is the probability that the number of fluid ounces dispensed is between 11.7 and 12.2?
8. What is the probability that the number of fluid ounces dispensed is less than 12.1?
9. What is the probability that the number of fluid ounces dispensed is more than 12.4?
10. What is the probability that the number of fluid ounces dispensed is less than 11.65 or more than 12.35?

The waiting time for a teller at the North Side Community Bank on a Friday night is normally distributed with a mean of 3.8 min and a standard deviation of 0.4 min. The bank manager is concerned with customer satisfaction and has initiated a policy of giving a $5 gift card to any customer who has to wait longer than 4.5 min to be served. Use this information to answer question 11.

11. What is the probability that a Friday night customer will receive a $5 gift card from the manager?

Regressions

Many students have the perception that mathematics is not real because the answers to the equations done in class are always "nice" numbers (integers, terminating decimals, or fractions with reasonable denominators). The reality is that the entire world uses mathematics, and the numbers we deal with, as well as the answers we get, are not always "nice." With the inclusion of technology in the classroom, you have had the opportunity to see more and more real applications of mathematics. It is often the case that data are gathered from some research, and in analyzing the data, a mathematical equation is sought to relate two or more variables so that predictions can be made about future behavior. The process through which the equation is found is called **regression.**

PROBLEM A diver needs to be aware of the pressure that is being applied to his/her body and equipment when swimming in the ocean. The following table shows pressure data (in atmospheres) for different depths (in meters).

DEPTH	PRESSURE
10	1.99
15	2.49
20	2.98
22	3.18
25	3.48
35	4.47

a. Make a scatterplot of the data with depth as the independent variable.

b. Determine the type of relationship that appears to be present between these variables.

c. Determine the equation of best fit for the data.

d. Predict the amount of pressure on a diver at a depth of 27 m.

SOLUTION a. The scatterplot for the data is shown in the figure.

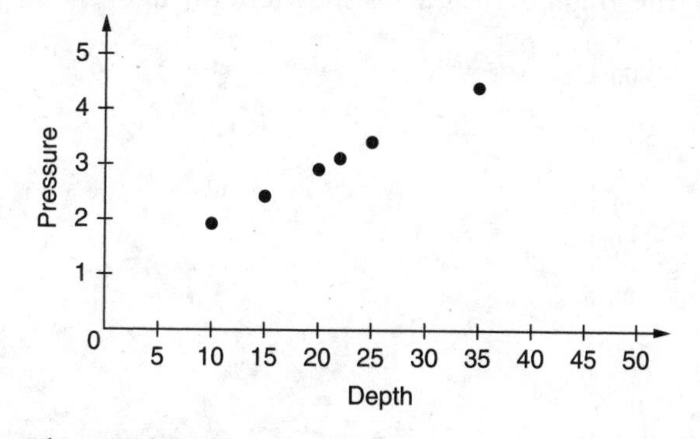

b. The graph indicates that there is a linear relationship between the variables.

c. Use the linear regression tool of your calculator or computer program to determine that the equation of best fit is pressure $= 0.099 \times$ depth $+ 1$.

d. The slope of the line is approximately 0.1 atm/m, while the vertical intercept is 1 atm. The vertical intercept indicates that the pressure on the surface of the water is 1 atm, and the slope shows that the pressure will increase by about 1 atm for every 10 m of depth.

According to this model, the pressure on a diver at a depth of 27 m will be $0.099 \times 27 + 1$ or 3.68 atm. This calculation was used from the regression equation within the calculator, and not the equation of best fit, which has the parameters rounded.

In addition to linear regressions, you should also be familiar with **exponential regressions** (equations of the form $y = a \times b^x$), **power regressions** (equations of the form $y = a \times x^b$), and **logarithmic regressions** (equations of the form $y = a + b\ln(x)$).

PROBLEM The data provided in the following table represent the stopping distance (in feet) when a vehicle on a dry road and with good tread on its tires is being driven at the given speed (in miles per hour).

SPEED	DISTANCE
30	37.6
35	51.2
40	66.9
45	84.7
50	104.5
55	126.5
60	150.5
65	176.7
70	204.9
80	267.6

a. Make a scatterplot for the data.

b. Determine the type of relationship that appears to be present between the variables.

c. Determine the equation of best fit for the data.

d. If the vehicle and conditions meet the requirements for the data given, predict the stopping distance in feet when the vehicle is traveling at a speed of 58 miles per hour.

SOLUTION a. The graph of the data is shown in the figure.

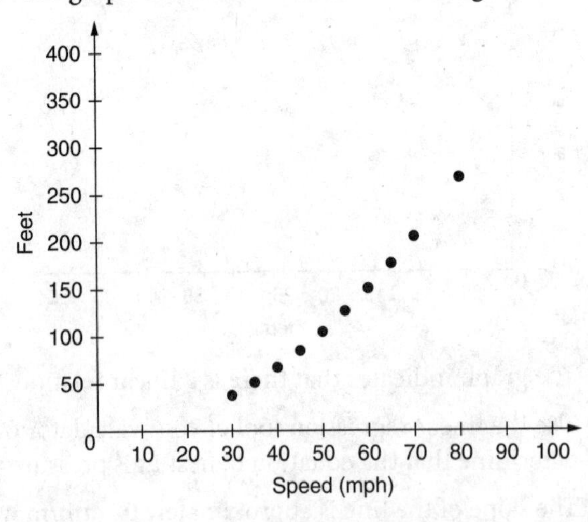

b. The curvature of the graph might indicate that the relationship is exponential, or it may be polynomial. A logarithmic relationship is unlikely because the graph is increasing.

c. Use your calculator or computer to determine the equation of best fit using an exponential regression and a power regression.

	EQUATION	r VALUE
Exponential	$\text{dist} = 13.691(1.0400)^{\text{speed}}$	0.98991
Power	$\text{dist} = 0.042(\text{speed})^{2.001}$	1

The power regression returns an r value of 1. The r value, called the **Pearson correlation coefficient**, gives a measure of the strength of the equation. It will always be the case that $-1 \leq r \leq 1$. The closer $|r|$ is to 1, the better the equation fits the data. This rule only applies to linear, power, exponential, and logarithmic regressions. The reason for this is that each of these curve forms can be linearized through the use of logarithms.

d. It appears that the distance needed to stop a car on a dry road is approximately quadratic. The distance needed to stop a car traveling at 58 miles per hour is $\text{dist} = 0.042(58)^{2.001} = 141.9$ ft.

PROBLEM The height (in meters) of the bouncing of a ball is measured with a CBL. The data are filtered to find the maximum height of the ball after each bounce. The data are shown in the accompanying table.

BOUNCE	HEIGHT (M)
1	1.073
2	0.856
3	0.722
4	0.572
5	0.498
6	0.431

a. Make a scatterplot for the data.

b. Determine the type of relationship that appears to be present between the variables.

c. Determine the equation of best fit for the data.

d. Extrapolate from this model to predict the height of the ball after the seventh bounce.

SOLUTION a. The scatterplot for these data is shown in the illustration.

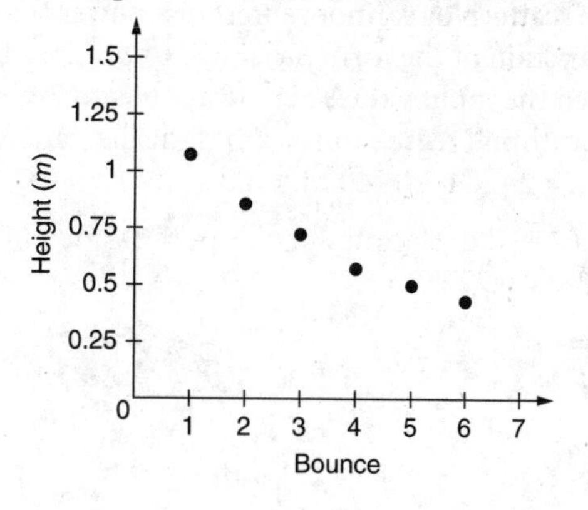

b. The scatterplot indicates that the data fit an exponential model.

c. Use your graphing calculator or computer to determine that the equation of best fit is height $= 1.251(0.832)^{\text{bounce}}$ ($r = -0.99605$).

d. The height of the ball after the seventh bounce will be $1.251(0.832)^7 = 0.345$ m.

PROBLEM The following table gives the wind chill factors when the air temperature is 20°F and the wind is blowing at the specified speed (in miles per hour).

WIND SPEED (MPH)	WIND CHILL
5	13
10	9
15	6
20	4
25	3
30	1
40	–1

a. Make a scatterplot for the data.

b. Determine the type of relationship that appears to be present between the variables.

c. Determine the equation of best fit for the data.

d. Predict the wind chill when the wind is blowing at 35 miles per hour.

SOLUTION a. The scatterplot for the data is shown in the figure.

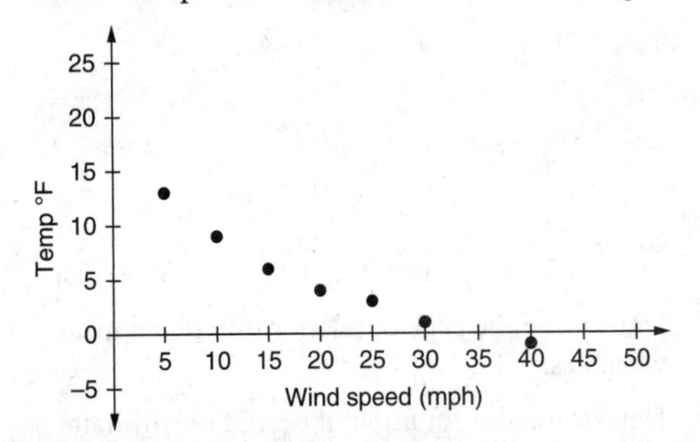

b. The curvature of the scatterplot shows that a linear function is unlikely and the presence of both positive and negative values eliminates exponential functions as a possibility.

c. The scatterplot cannot represent a power function because it is impossible for a function of the form $f(x) = ax^b$ to produce both positive and negative outputs when the input values are always positive. Consequently, it appears that a logarithmic regression will best fit the data. The equation of best fit is wind chill $= 24.184 - 6.740 \ln(\text{wind speed})$ ($r = -0.997597$).

d. At a speed of 35 miles per hour, the wind chill will be $24.184 - 6.740 \ln(35) = 0.22$°F.

EXERCISE

10·4

Solve the following.

1. Real estate ads for a metropolitan area reveal the following data for the relationship between the number of square feet of living space and the asking price for condominiums.

AREA	PRICE	AREA	PRICE	AREA	PRICE
1200	$380,000	1500	$440,000	1800	$550,000
1250	$375,000	1600	$470,000	2000	$650,000
1300	$390,000	1600	$485,000	2100	$800,000
1400	$410,000	1700	$490,000	2200	$850,000
1450	$425,000	1750	$500,000	2300	$950,000

   a. Make a sketch of the price for a condominium in terms of the number of square feet of living space.

   b. Determine if there is an exponential relationship between the amount of living area and the price of a condominium.

   c. Write an equation for the price of a condominium in terms of the number of square feet of living area.

   d. Explain the meaning of the base of the exponential statement for this equation.

   e. How much should one expect to pay for a condominium with 1900 sq ft of living area?

2. The data in the following table show the number of grams of fat and the number of calories from fat in McDonald's sandwiches.

SANDWICHES	CALORIES FROM FAT	TOTAL FAT
Hamburger	80	9
Cheeseburger	110	12
Double Cheeseburger	210	23
Quarter Pounder®	170	19
Quarter Pounder® with Cheese	230	26
Double Quarter Pounder® with Cheese	380	42
Big Mac®	260	29
Big N' Tasty®	220	24
Big N' Tasty® with Cheese	250	28
Filet-O-Fish®	160	18
McChicken®	150	16
Premium Grilled Chicken Classic Sandwich	90	10
Premium Crispy Chicken Classic Sandwich	150	17
Premium Grilled Chicken Club Sandwich	190	21
Premium Crispy Chicken Club Sandwich	250	28
Premium Grilled Chicken Ranch BLT Sandwich	140	16

(continued)

SANDWICHES	CALORIES FROM FAT	TOTAL FAT
Premium Crispy Chicken Ranch BLT Sandwich	200	23
Snack Wrap™ with Ranch	140	16
Grilled Snack Wrap™ with Ranch	90	10
Snack Wrap™ with Honey Mustard	130	15
Grilled Snack Wrap™ with Honey Mustard	80	9

Data from http://nutrition.mcdonalds.com/nutritionexchange/nutritionfacts.pdf

a. Sketch a graph of the number of calories from fat in a McDonald's sandwich in terms of the number of grams of fat in the sandwich.

b. Determine if there is a linear relationship between the number of calories from fat in a McDonald's sandwich and the number of grams of fat.

c. Determine the equation for the line of best fit for the number of calories from fat in a McDonald's sandwich in terms of the number of grams of fat.

d. Explain the meaning of slope for this equation.

e. Predict the number of calories from fat in a McDonald's sandwich if the sandwich contains 20 grams of fat.

3. The price of Apple stock (in dollars) on the first trading day of August for the years 2001–2005 and 2007–2011 is listed in the following table.

YEAR	PRICE
2011	$373.72
2010	$243.10
2009	$168.21
2008	$169.53
2007	$138.48
2005	$46.89
2004	$17.25
2003	$11.31
2002	$7.38
2001	$9.27

Data from http://finance.yahoo.com/q/hp?s=AAPL

a. Make a scatterplot for the data, measuring time as the number of years since 2000.

b. Determine the type of relationship that appears to be present between the variables.

c. Determine the equation of best fit for the data.

d. Predict the price of Apple stock in 2006.

4. A model used for explaining radioactive decay produces the following data.

TIME (SEC)	NUMBER RADIOACTIVE NUCLEI
0.1	961
0.2	915
0.3	864
0.6	735
0.7	697
1.0	607

a. Make a scatterplot for the data.

b. Determine the type of relationship that appears to be present between the variables.

c. Determine the equation of best fit for the data.

d. Predict the number of radioactive nuclei after 0.5 sec.

5. The data in the table below represent the stopping distance (in feet) when a vehicle on a wet road and with poor tread on its tires is driving at the given speed (in miles per hour).

SPEED	DISTANCE
30	100.4
35	136.6
40	178.4
45	225.8
50	278.8
55	337.3
60	401.4
65	471.1
70	546.4
80	713.6

a. Make a scatterplot for the data.

b. Determine the type of relationship that appears to be present between the variables.

c. Determine the equation of best fit for the data.

d. Predict the stopping distance on a wet road with tires having poor tread from a speed of 58 miles per hour.

6. An astronomical unit (au) is defined to be the average distance from the center of the sun to the center of the Earth. The following table contains the number of astronomical units the planets in our solar system are from the sun and the number of Earth years it takes for each planet to make a revolution around the sun.

PLANET	DISTANCE FROM THE SUN	TIME FOR ONE REVOLUTION AROUND THE SUN
Mercury	0.39	0.24
Venus	0.72	0.62
Earth	1	1
Mars	1.52	1.88
Jupiter	5.2	11.86
Saturn	9.54	29.46
Uranus	19.18	84.01
Neptune	30.06	164.8

a. Make a scatterplot for the data.

b. Determine the type of relationship that appears to be present between the variables.

c. Determine the equation of best fit for the data.

d. Predict the time needed for the planetoid Ceres, which is 2.7 au from the sun, to make a complete revolution around the sun.

7. The following table gives the wind chill factors when the air temperature is 10°F.

WIND SPEED	WIND CHILL
5	1
10	–4
15	–7
20	–9
25	–11
30	–12
40	–15

a. Make a scatterplot for the data.

b. Determine the type of relationship that appears to be present between the variables.

c. Determine the equation of best fit for the data.

d. Predict the wind chill at 10°F when the wind is blowing at 35 miles per hour.

Inferential statistics

Whereas descriptive statistics yield measures of center and spread, **inferential statistics** allow us to get estimates for the parameters measuring center and proportion by using the corresponding sample statistics. They also provide a means for us to test the validity of claims about center and proportion. Although the study of inferential statistics covers a wide variety of scenarios, we will limit our discussion to large sample sizes. **Simulation** is a technique of using probability to model the sampling process. The importance of this is that simulation is faster and cheaper. Imagine, for example, that you want to test the breaking strength of a rivet or shatterproof glass. Because it costs money to break these items and then they cannot be used again, a probabilistic model is employed.

Basics

A key issue when attempting to measure some aspect of the population through sampling is randomization. Designing surveys and sampling processes is very tricky work because one wants to avoid any type of bias when collecting data. For example, if members of the high school dance committee want to determine the type of music those attending the dance would want to have, asking the members of only one homeroom might not be indicative of the entire group. Notice that it is not necessarily biased, but the process does leave open that possibility.

PROBLEM The members of the high school dance committee want to determine the type of music those attending the dance would want to have. State three ways that the committee can randomly select students to survey.

SOLUTION 1: The names of all the school's students can be accessed through a database. Use a random number generator to pick names based on their position within the list.

2: Randomly select a set of homerooms from a list of all the homerooms in the school and ask the students in each homeroom to respond to the survey.

3: Separate the students by their grade level and then randomly choose (by either method 1 or 2 earlier) equal numbers of students from each grade level.

These are three possible solutions. Many others can be used.

1. A news service is trying to predict the outcome of an upcoming election. They randomly select 1000 people who live in the northeastern part of the United States. Does this represent an unbiased survey? Explain.
2. A news service is trying to predict the outcome of an upcoming election. Describe two processes that could be used to collect an unbiased set of data.

Central limit theorem and standard error

As we saw in Chapter 10, the normal distribution enables us to compute the probability that a piece of data will lie within an interval, provided we know the mean and standard deviation of the distribution. Although not all distributions are normally distributed (think of a single fair die, the probability for each outcome 1 through 6 is the same, one-sixth, making the distribution uniform), there is a key theorem in statistics that allows us to apply the normal distribution while attempting to make inferences about the center of the distribution.

The **Central Limit Theorem** states that as the sample size gets larger, the sampling distribution of the mean can be approximated by the normal distribution. This is true whether or not the original distribution is unimodal and symmetric. That is to say, if we take repeated samples from a population, compute the mean of the sample, and analyze the distribution of the means, this distribution will be approximately normally distributed, provided the size of each sample is sufficiently large.

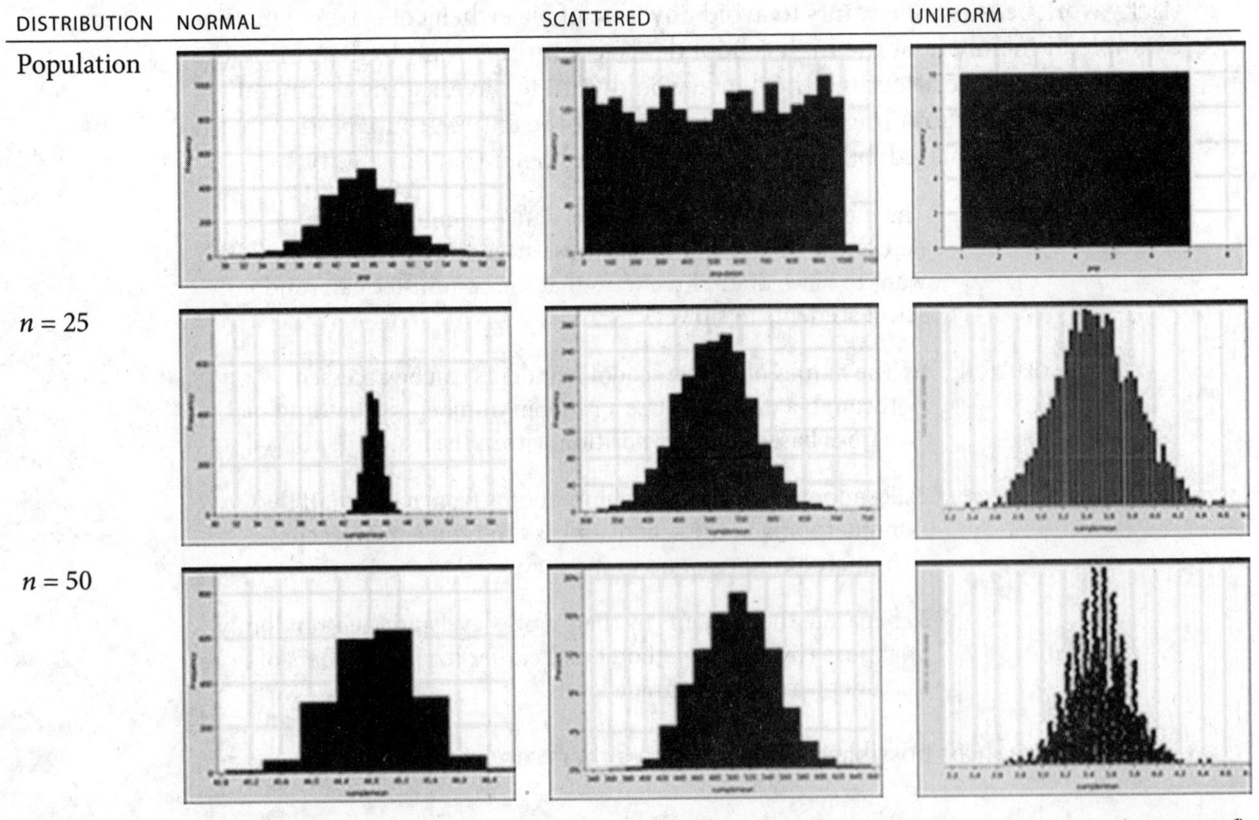

(continued)

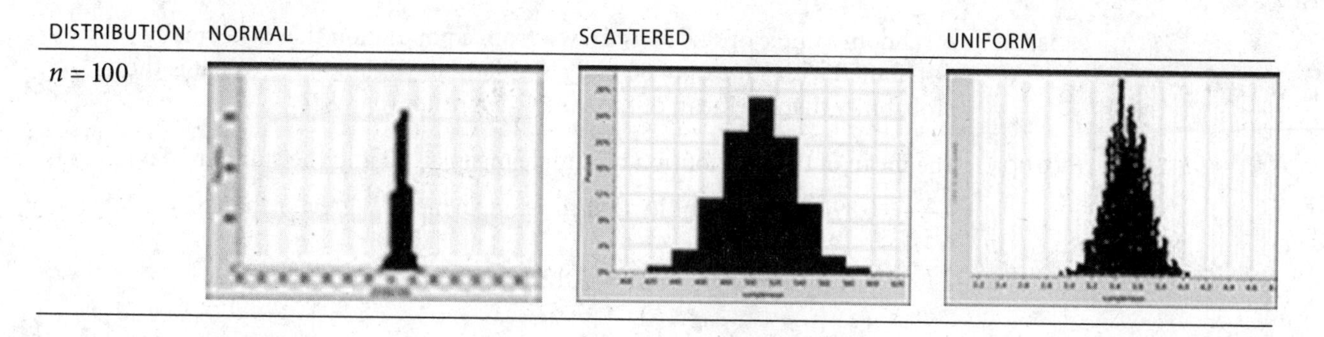

Two key pieces to this theorem are that given a population with mean, u, the mean of all the sample means, $u_{\bar{x}}$, is the same value and the standard deviation of the means (called the **standard error of the means**) is $\sigma_{\bar{x}} = \frac{\sigma}{\sqrt{n}}$.

We can then apply the normal distribution to compute probabilities for the mean of the selected sample. This is the basis for the work on inferential statistics involving the parameters (mean and proportion) for large samples. (There is a great deal more to the topic of inferential statistics that is well beyond the scope of this course. The intent at this level is to give the student a feel for how decisions are made based on the use of statistics.)

PROBLEM A large number of samples of size 36 are drawn from a population with mean 42 and standard deviation of 1.8. What is the mean and standard error for these sample means?

SOLUTION The mean remains as 42, but the standard error of the mean is $\frac{1.8}{\sqrt{36}}=0.3$.

PROBLEM A random sample of size 36 is drawn from a population with mean 42 and standard deviation of 1.8. What is the probability that the mean of the selected sample is between 41.5 and 42.5?

SOLUTION Using the results of the previous problem, the standard error of the mean is 0.3. The probability that the mean is between 41.5 and 42.5 is $P(41.5 < \bar{x} < 42.5) =$ 0.9044. (Use your graphing calculator to find this value.)

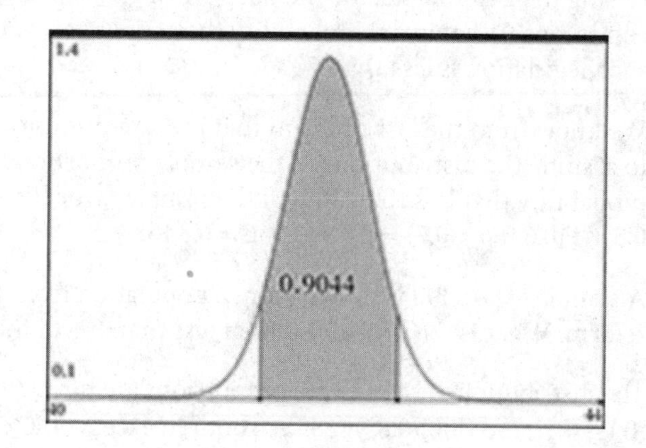

PROBLEM A random sample of size 50 is drawn from a population that is normally distributed with a mean of 80 and a standard deviation of 6.3. What is the probability that the mean of the sample is greater than 82?

SOLUTION The mean of the distribution of sample means is 82 and the standard deviation is $\frac{6.3}{\sqrt{50}} = 0.891$. The probability that the sample mean is greater than 82 is $P(\bar{x} > 82) = 1 - P(0 < \bar{x} < 82) = 0.012$.

The sampling distribution for proportions, p, is similar to that of the mean. The distribution of the sample proportions can be approximated by a normal distribution with sample size n if the product $np \geq 10$ and $n(1 - p) \geq 10$. The mean of the proportions will be p, and the standard error is equal to $\sqrt{\frac{p(1-p)}{n}}$.

PROBLEM Repeated samples of size 25 are drawn from a population for which 60% are in favor of raising the gasoline tax for the purpose of gaining revenue to improve road conditions. What is the mean and standard error for these samples?

SOLUTION The mean proportion for the samples is that of the population, 0.6, and the standard error for the proportion is $\sqrt{\frac{(0.6)(0.4)}{25}} = 0.098$.

PROBLEM A sample of size 25 is drawn from a population for which 60% are in favor of raising the gasoline tax for the purpose of gaining revenue to improve road conditions. What is the probability the proportion of those selected in favor of such legislation is greater than 55%?

SOLUTION We can assume that the distribution of the proportions is normally distributed with proportion 0.6 and standard error 0.098 because (0.6)(25) = 15 and (0.4)(25) = 10 and therefore meet the criteria for assuming a normal distribution. The probability that more than 55% of the sample voters favor the legislation is $P(p > 0.55) = 1 = P(0 < p < 0.55) = 0.695$.

PROBLEM A sample of size 25 is drawn from a population for which 60% are in favor of raising the gasoline tax for the purpose of gaining revenue to improve road conditions. What is the probability the proportion of those selected in favor of such legislation is less than 70%?

SOLUTION We know from the last problem that the conditions of the problem allow us to assume the distributions of the sample proportions will be normal. The probability that less than 70% of the sample favor the legislation is $P(p < 0.7) = 0.5 + P(0.6 < p < 0.7) = 0.5 + 0.346 = 0.846$.

PROBLEM A sample of size 100 is drawn from a population from which 15% oppose tax reform. What is the probability that less than 8% of the sample oppose tax reform?

SOLUTION The distribution of the sample proportions are normally distributed because (100)(0.15) = 15 and (100)(0.85) = 85 are both in excess of 10. The mean proportion for the sampling distribution is 0.15, and the standard error is $\sqrt{\frac{(0.15)(0.85)}{100}} = 0.0357$. Therefore, $P(p < 0.08) = 0.5 - P(0.08 < p < 0.15) = 0.5 - 0.475 = 0.025$.

For questions 1–4, determine the standard error for the parameter indicated in each problem.

1. Standard deviation = 5.7; sample size = 36
2. Standard deviation = 12.8; sample size = 100
3. Proportion = 0.7; sample size = 125
4. Proportion = 0.2; sample size = 500

Use the following example to answer questions 5–7.

The height of the students at Niskayuna High School has a mean of 67.2 inches with a standard deviation of 3.4 inches. A random sample of 50 students is selected and their heights measured.

5. What is the probability that the mean height of the students is between 66.5 and 67.9 inches?
6. What is the probability that the mean height of the students is less than 68.5 inches?
7. What is the probability that the mean height of the students is greater than 66 inches?

Use the following example to answer questions 8–10.

Ninety-eight percent of all the batteries made at the Fort Mill factory meet the manufacturer's specifications. A random sample of 500 batteries is selected for testing.

8. What is the probability that between 97.5% and 99% of the batteries meet the manufacturer's specifications?
9. What is the probability that at least 99.5% of the batteries meet the manufacturer's specifications?
10. What is the probability that less than 1% of the batteries **fail** to meet the manufacturer's specifications?

Standardized (*z*) scores

Mrs. Stockwell is teaching three sections of Algebra II. She gives different versions of an exam on statistics to the three sections. John, Alice, and Julio, three friends who are each in different sections, meet after the exams are returned and are surprised to see that they each scored 90% on the test. John commented that he was really pleased with his result because the average grade on the exam in his class was 84. Alice also expressed her pleasure at her score because the average grade in her class was 86. Julio seemed a little bummed by this conversation. He explained that the average score in his class was 92. (Mrs. Stockwell confirmed that all three versions were comparable in their level of difficulty and that the scores on all three exams had a standard deviation of 4 points.) As people will often do, the three friends discuss which of them did better. John argued that he did the best because his score was 1.5 standard deviations higher than the mean, whereas Alice scored 1 standard deviation above the mean and Julio scored a half a standard deviation below the mean.

Although you can argue that all three did pretty well scoring a 90 on their exam, the argument presented by John is a valid method of comparing values from different populations. Determining the number of standard deviations from the mean where a data point is located is called a **standardized score** and, as it is traditionally represented by the variable z, is also called a

z-score. (Before the availability of computing devices to compute probabilities under the normal curve, the values were computed from a table of the standard normal distribution. This distribution had a mean of 0 and a standard deviation of 1. Students first computed the z-score from the raw data and then used the table to read the given probabilities.)

You know that the probability a data value from a variable that is normally distributed will lie within one standard deviation of the mean is approximately 68%. Said differently, $P(-1 < z < 1) = 68\%$.

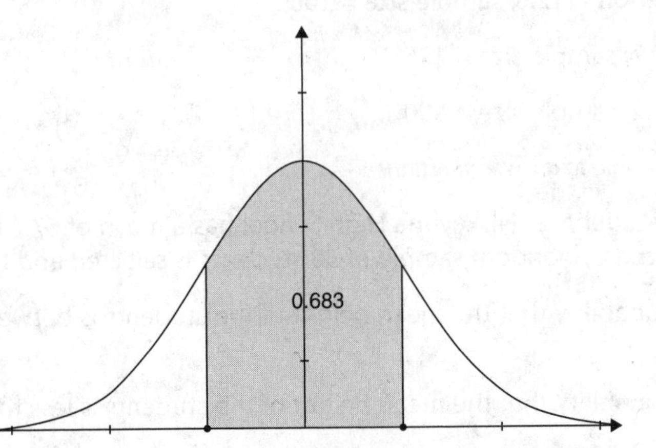

Between which two standard scores, symmetric about the mean, will 50% of the data lie? Rather than go through a great deal of trial and error, we will take advantage of the invNorm function that is available on the calculator. The structure of the command for this function is invNorm(Area, Mean, Standard Deviation). It is important to note that the Area input represents the area under the normal curve to the left of the value we seek.

To determine the value of A in the diagram, realize that there is 25% of the area to the left of A. Therefore, A = invNorm(0.25, 0, 1) = −0.67449. You can easily tell that the value of B must be 0.67449 because the region is symmetric about the mean, 0.

Suppose the distribution in question has a mean of 149.3 and a standard deviation of 18.2. The value of A is invNorm(0.25, 149.3, 18.2) = 137.024. You can still use symmetry to find B, but you might find it easier to replace the 0.25 for the Area input used to find A with 0.75. B = invNorm(0.75,149.3,18.2) = 161.576.

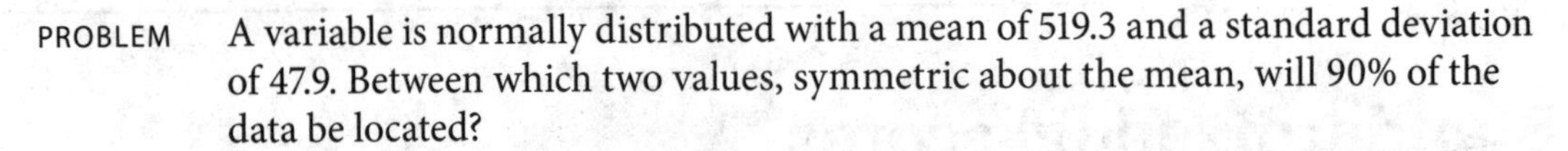

PROBLEM A variable is normally distributed with a mean of 519.3 and a standard deviation of 47.9. Between which two values, symmetric about the mean, will 90% of the data be located?

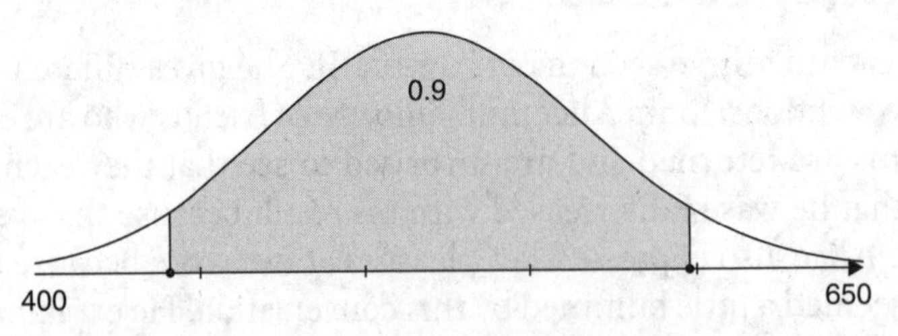

SOLUTION A = invNorm(0.05, 519.3, 47.9) = 440.512 and B = invNorm(0.95, 519.3, 47.9) = 598.088

PROBLEM The height of students at Central High is normally distributed with a mean of 68.7 inches and a standard deviation of 4.1 inches. What is the height that represents the point where only 2.5% of the students are taller?

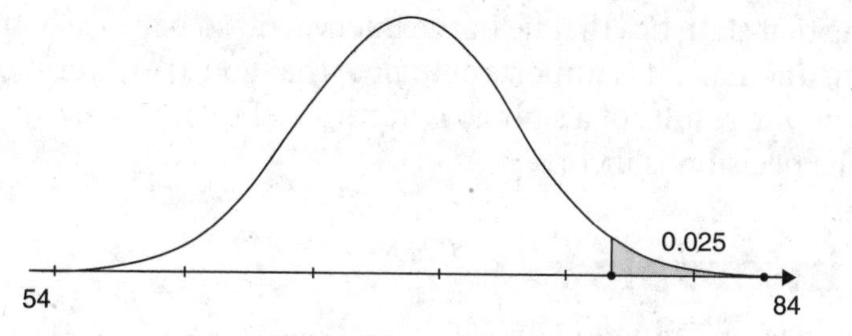

SOLUTION If 2.5% of the area is to the right of point X, then 97.5% must be to the left. Therefore, X = invNorm(0.975, 68.7, 4.1) = 76.74 inches.

EXERCISE

11·3

For problems 1–3, determine the z-score for the given value of x.

1. $x = 25$, mean = 18, standard deviation = 4
2. $x = 25$, mean = 28, standard deviation = 4
3. $x = 25$, mean = 21, standard deviation = 2.5

For problems 4 and 5, determine the values of A and B assuming that they are symmetric about the mean.

4. Mean = 87.2, standard deviation = 5.6, area between A and B is 80%
5. Mean = 904, standard deviation = 93.7, area between A and B is 60%
6. The volume of soft drink in a 1-liter bottle of Monahan's Fabulous Elixirs is normally distributed with a mean of 1.002 liters and a standard deviation of 0.03 liters. Ninety-nine percent of the bottles contain more than *X* liters. What is the value of *X*?
7. Mrs. Netoskie rides hunters in equestrian competitions. The heights of these hunters are normally distributed with a mean of 16 hands and a standard deviation of 1.2 hands. What is the height at which only 4% of the horses are taller?

The basics of inferential statistics

As was noted earlier in the chapter, statistical decisions are based on the study of probability. A fair die is one in which the chances of getting heads or tails are equally likely. If someone flipped a coin 10 times and it came up heads each time, might you suspect that the coin is not fair? One might consider this to be the case. However, keep in mind that probability models are based on a very large number of repetitions, and 10 is not a very large number. If the coin showed heads on 9000 out of 10,000 flips, you might have a concern (and hopefully you thought that 10,000 is not that large a number either).

So, how does one make a decision based on probability? In the case of discrete cases (experiments in which one can count the number of successes), it is necessary that the Law of Large Numbers is applied. If you have enough examples, you can make an educated guess as to what the reality is. In the case of **continuous data** (data that is measured rather than counted), we rely on the normal curve. (Again, our concentration at this level is on the normal curve and the measure of means and proportions.) Considering that more than 99% of the data associated with a variable that is normally distributed will lie within three standard deviations from the mean, it is safe to assume that statistics that lie in the interval do so because of natural variability rather than because of an anomaly. It is important to note that you are never 100% certain that the decision made based on the results of a sample is accurate. The best one can hope for is to minimize the chance that the decision is incorrect.

Confidence intervals

Let's begin with the situation that the population parameter for a continuous variable is unknown and the purpose of the sample is to get an estimate of its value. Using the Central Limit Theorem, we know that so long as we choose a large enough sample (curiously, 30 turns out to be the magic number for a sufficiently large sample size), the distribution of the corresponding statistic (mean or proportion) will be normally distributed. (You'll find that as you study more about statistics that authors will start with the case that we assume we do not know the population mean but do know the population standard deviation. That seems like a bit of a reach, so we'll work with the case that neither is known but will use the sample standard deviation in our calculations.) We acknowledge that our response is an estimate by giving a measure of the level of confidence we have with our result (and hence the term **Confidence Interval**).

PROBLEM A random sample of 100 banking customers who were at a particular branch during the lunch hour is selected. The time they waited while in line before a teller helped them is measured. The mean of the data is 4.7 minutes with a standard deviation of 1.3 minutes. Determine an interval for the mean waiting time of a lunch-time customer at this branch using a 95% level of confidence.

SOLUTION We begin by looking at a standard normal curve (remember—the mean is 0 and the standard deviation is 1). For what two values will 95% of the data lie between? A = invNorm(.025,0,1) = −1.96 and B = invNorm(.975,0,1) = 1.96.

We know that z-scores are computed with the formula $z=\dfrac{x-\mu}{\sigma}$. In this case, the data value is the mean of the sample, and the spread is actually the standard error of the mean, so the equation we need will be $z=\dfrac{\overline{x}-\mu}{\dfrac{\sigma}{\sqrt{n}}}$. Because we are using the sample standard deviation, the equation is $z=\dfrac{\overline{x}-\mu}{\dfrac{\sigma}{\sqrt{n}}}$. Solve this equation for μ, $\mu=\overline{x}-z\left(\dfrac{\sigma}{\sqrt{n}}\right)$. Because the z-score will always be + the same value, the formula for computing the endpoints of the confidence interval becomes $\mu=\overline{x}\pm z\left(\dfrac{\sigma}{\sqrt{n}}\right)$.

In the case of the waiting time for the customers at the bank, the mean waiting time is $4.7-(1.96)\left(\dfrac{1.3}{\sqrt{100}}\right)=4.445$ minutes to $4.7+(1.96)\left(\dfrac{1.3}{\sqrt{100}}\right)=$ 4.955 minutes. (With this result, the manager could advertise that customers are served in less than 5 minutes while on their lunch break!)

PROBLEM A random sample of 75 students from Hilltop High School had their heights measured. The average for the sample was 68.2 inches with a standard deviation of 2.5 inches. Determine an interval for the true mean height of the students at Hilltop High using a 90% level of confidence.

SOLUTION The z-score associated with a 90% level of confidence is 1.645. Therefore, the endpoints for the confidence interval are $68.2 \pm (1.645)\left(\frac{2.5}{\sqrt{75}}\right)$ which equals 67.7 and 68.7. There is a 90% chance that the true mean height of the Hilltop High School students is between 67.7 and 68.7 inches.

Your graphing calculator is able to do all this work for you. Using the TI-Nspire, go to a calculator page. Use the Statistics tool (#6), confidence intervals (#6), and z-interval (#1); in this case data input is from Stats. The standard deviation is 2.5, the mean is 68.2 (yes, use the sample statistics even though the menu identifies the parameters), n is 75, and change the C-level to 0.90.

zInterval 2.5,68.2,75,0.9: *stat.results*

"Title"	"z Interval"
"CLower"	67.7252
"CUpper"	68.6748
"$\bar{x}$"	68.2
"ME"	0.474828
"n"	75.
"σ"	2.5

The value ME, mean error, is the value of $z\left(\frac{\sigma}{\sqrt{n}}\right)$.

Using the TI-84, press the Stat key, slide right to tests, choose option 7 ZInterval, enter the values, and press Enter when the cursor is on calculate.

```
NORMAL FLOAT AUTO REAL RADIAN MP
EDIT CALC TESTS
1:Z-Test...
2:T-Test...
3:2-SampZTest...
4:2-SampTTest...
5:1-PropZTest...
6:2-PropZTest...
7:ZInterval...
8:TInterval...
9↓2-SampZInt...
```

```
NORMAL FLOAT AUTO REAL RADIAN MP
ZInterval
Inpt:Data Stats
σ:2.5
x̄:68.2
n:75
C-Level:0.9
Calculate
```

```
NORMAL FLOAT AUTO REAL RADIAN MP
ZInterval
(67.725,68.675)
x̄=68.2
n=75
```

PROBLEM A survey of 150 citizens of voting age is taken in a town to determine if there is support for a bill proposing new recycling regulations. Seventy percent of those surveyed indicated that they supported the bill. Determine an interval for the proportion of the population that supports the bill at the 98% level of confidence.

SOLUTION With $p = 0.7$, the standard error of the proportion is $\sqrt{\frac{(0.7)(0.3)}{150}} = 0.037$. The z-score for a 98% confidence interval is 2.326. Therefore, the endpoints of the confidence interval are $0.7 \pm (2.326)(0.037)$, which equal 0.614 and 0.786.

Using the Nspire to determine this result requires that you give the number of successes (rather than the percentage). Seventy percent of 150 is 105, n is 150, and the C-level is 0.98.

zInterval_1 Prop 105,150,0.98: *stat.results*

"Title"	"1–Prop z Interval"
"CLower"	0.612956
"CUpper"	0.787044
"$\hat{p}$"	0.7
"ME"	0.087044
"n"	150.

The town officers can be 98% confident that between 61.4% and 78.6% of the voting citizens favor the bill.

Using the TI-84, press the Stat button, slide to Tests, and option A is 1-PropZInt. Enter 105 and 150 for the results of the survey and set C-Level to 0.98.

1. The personnel department of a corporation wants to estimate the average amount of money spent annually on dental expenses. The results of a random sample of 50 employees show an average annual expense of $352.18 with a standard deviation of $50.30. Determine a 90% confidence interval for the average amount of money spent annually on dental expenses.

2. The manager of a rock-and-roll band wants to estimate the mean number of people who attend a concert put on by her clients. A random sample of the band's last 40 concerts shows an average attendance of 11,700 people with a standard deviation of 1200. Determine an interval for the average attendance at the 98% level of confidence.

3. An automobile dealership manager wants to determine the proportion of new car transactions that have the customer select a lease option rather than purchase. The manager randomly selects 60 records and determines that 45% of all transactions involve a lease option. Determine an interval for the proportion of monthly transactions on new cars that involve a lease option at the 95% level of confidence.

Tests of hypotheses

Your friend claims, "The average number of minutes of music played by this radio station each hour is 40 minutes."

You respond, "I disagree."

What are the implications of the statement "I disagree"? Do you believe that the station is playing less than 40 minutes per hour, more than 40 minutes per hour, or just that it is not 40 minutes of music per hour on average? When performing a statistical test of a hypothesis, it is important to know which condition you are claiming as an alternative. Here's why.

Without examining the entire population under consideration, there will always be the possibility that the claim being made is accurate and the alternative is incorrect. The goal of the test of hypothesis is to make the probability that this happens as small as is reasonably acceptable. A classic argument comes from our legal system. The charge is that Colonel Plum killed Mr. X. The operating tenet in the legal system is "innocent until proven guilty." The prosecutors present their evidence in an attempt to show beyond reasonable doubt that Colonel Plum is guilty. The defense lawyer, if unable to out and out prove the innocence of the defendant, tries to present evidence to show reasonable doubt. When both sides are done, the case goes to the jury for deliberation. Due to the premise of innocent until proven guilty, the initial hypothesis is that Colonel Plum is innocent. There are four possible scenarios:

A: Colonel Plum is innocent and the jury finds him guilty. An incorrect decision is made.
B: Colonel Plum is innocent and the jury acquits him. A correct decision is made.
C: Colonel Plum is not innocent and the jury acquits him. An incorrect decision is made.
D: Colonel Plum is not innocent and the jury finds him guilty. A correct decision is made.

Of the two incorrect decisions, scenario A is considered to be the more serious because our society never wants to put an innocent person in jail. We try to make the size of "reasonable doubt" small.

In the language of statistics, the initial hypothesis is called the **null hypothesis**, often designated as H_0. Competing with this is the **alternative hypothesis**, H_a. The amount of reasonable doubt is called the **Level of Significance**, often represented by α. As we did with our study of confidence intervals, we are only going to consider scenarios that lead themselves to using the normal distribution. Reasonable doubt is any case in which the offered alternative can be explained as a reasonable deviation from a stated mean or proportion.

Let's go back to the discussion of the amount of music being played by the radio station and examine the three alternatives we discussed.

Case 1

H_0: The average amount of music played per hour is not 40 minutes.
H_a: The average amount of music played per hour is less than 40 minutes.

The graph for the problem is shown next. The shaded region represents the area of reasonable doubt, and the unshaded region represents an area where natural variation from the mean is not reasonable. (Please note that if the station is actually playing music for more than 40 minutes, the null hypothesis is still accepted.)

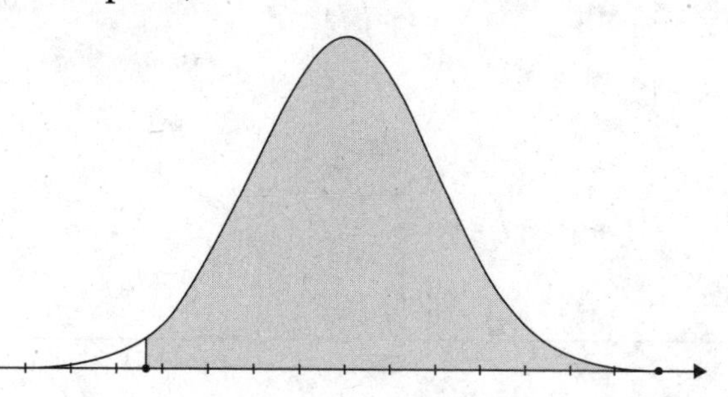

Case 2

H_0: The average amount of music played per hour is not 40 minutes.
H_a: The average amount of music played per hour is greater than 40 minutes.
The graph for the problem is shown here. The shaded region represents the area of reasonable doubt, and the unshaded region represents an area where natural variation from the mean is not reasonable. (Please note that if the station is actually playing music for less than 40 minutes, the null hypothesis is still accepted.)

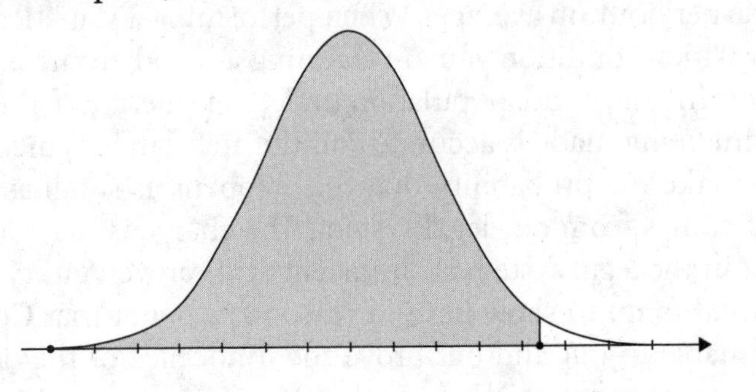

Case 3

H_0: The average amount of music played per hour is 40 minutes.
H_a: The average amount of music played per hour is not 40 minutes.
The graph for the problem is shown here. The shaded region represents the area of reasonable doubt, and the unshaded region represents an area where natural variation from the mean is not reasonable. Note that each unshaded tail contains one-half the value of α.

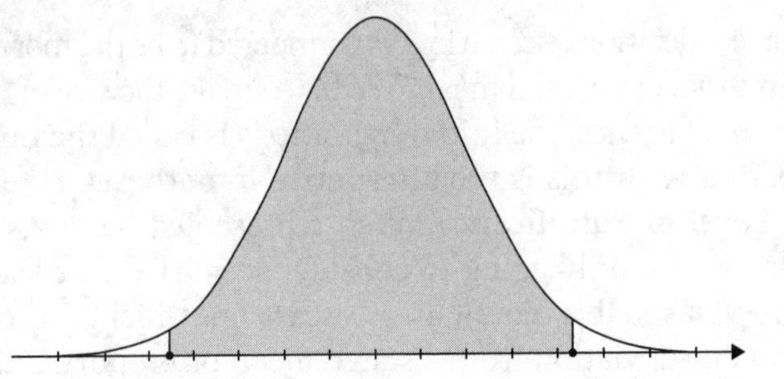

Suppose that your friend collects data from 50 randomly selected one-hour intervals and determines that the mean number of minutes of music played is 39.6 minutes with a standard deviation of 1.3 minutes. Let's put the level of significance at 5%. The area of the unshaded region is 0.05. We can determine the point on the normal curve for which this is true. We use the claimed measure of the mean, 40, to determine this point. The critical point, the point about which reasonable doubt is released, is $invNorm\left(0.05, 40, \frac{1.3}{\sqrt{50}}\right) = 36.6976$.

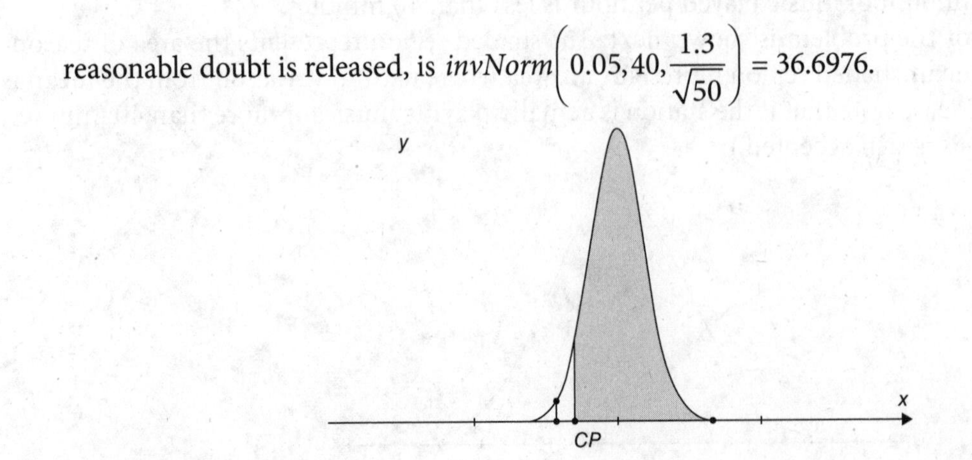

Because the sample mean is less than this value, it can be claimed that you are correct. The station is playing less than an average of 40 minutes of music per hour.

PROBLEM McBurger, Inc., claims that their half-pound burgers contain 48 g of fat. A random sample of 40 of their half-pound burgers contains an average of 50 g of fat with a standard deviation 0.8 grams. Does this data support the claim that the half-pound burgers at McBurger, Inc., have more than 48 g of fat? Use a 3% level of significance.

SOLUTION The null hypothesis is that the mean is 48, whereas the alternative hypothesis is that the mean is greater than 48. Because the alternative hypothesis claims the mean to be a larger value than that claimed by the manufacturer, the critical point will be at the right end of the normal distribution. The critical point for this test is $invNorm\left(0.97, 48, \frac{0.8}{\sqrt{40}}\right) = 48.23279$. Because the sample mean exceeds the critical point, it can be claimed that the mean number of grams of fat in a half-pound burger from McBurger, Inc., is greater than 48.

PROBLEM Conventional wisdom had been that 90% of the graduating class of Peakville County high schools went on to continue their education at a four-year college or university. A random sample of 36 graduating classes over the past few years shows that the proportion of graduates going to a four-year program is 87%. At the 5% level of significance, is this rate different from that of 90%?

SOLUTION The null hypothesis is that the proportion is 90%, and the alternative is that it is not. Without claiming to be higher or lower, this creates a two-tail test so that each tail on the bell curve will contain an area of 0.025. The left critical point is $invNorm\left(0.025, 0.9, \sqrt{\frac{(0.9)(0.1)}{36}}\right) = 0.802$, and the right critical point is

$invNorm\left(0.975, 0.9, \sqrt{\frac{(0.9)(0.1)}{36}}\right) = 0.9980$. With the sample proportion at 0.87, there does not appear to be enough evidence to claim that the proportion has changed.

PROBLEM The chancellor of a large city's school system claims that at least 70% of all the juniors in the city's high schools are enrolled in an Algebra II course. A random sample of 40 of the high schools shows that 67% of the juniors are enrolled in an Algebra II course. Does this result contradict the statement made by the chancellor? Test the claim at the 5% level of significance.

SOLUTION The null hypothesis is that the proportion is greater than or equal to 0.70, whereas the alternative hypothesis is that the proportion is less than 0.70. The critical point is $invNorm\left(0.05, 0.7, \sqrt{\frac{(0.7)(0.3)}{40}}\right) = 0.5808$. With the sample proportion well exceeding this critical value, there is not enough evidence to challenge the chancellor's claim.

EXERCISE
11·5

For each problem, state the null and alternative hypotheses, the value of the critical point, the decision that should be made, and an interpretation of the result.

1. A manufacturer claims that the mean weight of his 1-pound bag of flour is 1.01 pounds. A random sample of 50 bags of flour gives a mean weight of 0.99 pounds with a standard deviation of 0.02 pounds. Does the data provide significant results to refute the manufacturer's claim? Use the 5% level of significance.
2. An old commercial claimed, "Three out of five dentists recommend that you choose sugarless gum." Wondering if this is still a valid statement, a survey was taken of 400 randomly selected dentists around the country. The results show that 225 of the dentists recommend sugarless gum. Does the result of the survey indicate that there has been a change to the recommendation for using sugarless gum? Use the 3% level of significance.
3. Management claims that the average number of hours of overtime given each week to workers has increased in the past year. The mean number of overtime hours per week for the last three years has been 40.4 with a standard deviation of 3.8 hours. A random sample of 35 weeks from the last year shows a mean of 42.8 hours of overtime. Does the data support management's claim at the 2% level of significance?
4. Mike and Jack were listening to the radio while they were working. Mike claimed that at least half the songs in rock and roll were about love. Jack disagreed with this statement. After they decided how to define a song involving love, they kept a tally of the songs heard from the various radio stations and Internet services they were using—being careful not to include the same song twice in their survey. Of the 125 songs they heard, 48 were classified as being about love. Does this data support Jack's claim? Use the 1% level of significance.

Simulation

Simulation is a technique that uses a probabilistic model to test a theory. For example, a supposedly "fair" coin showed heads 10 times in a row when flipped. Does 10 times in a row give an indication that the coin might be biased? Rather than take the time to collect a large number of samples of 10 flips of a coin, it is possible to create a computer program to simulate these flips using a random number generator and have the program record the number of heads. Use the graphing calculator to generate a set of random numbers. Designate 0 to represent tails and 1 to represent heads. The command randint(0,1,10) will generate a set of 10 numbers, 0 or 1. Count the number of times one appears to represent the number of heads on that trial. A program was used to represent 2000 repetitions of this action.

The accompanying figure illustrates the outcome of 2000 flips of 10 fair coins. Four out of the 2000 outcomes contain 10 heads. We know the probability of getting 10 heads in 10 flips is fairly small, $\binom{10}{10}\left(\frac{1}{2}\right)^{10} = \frac{1}{1024}$, but it is not "impossible." The result of the simulation is that we do not have sufficient evidence to doubt the coin is fair.

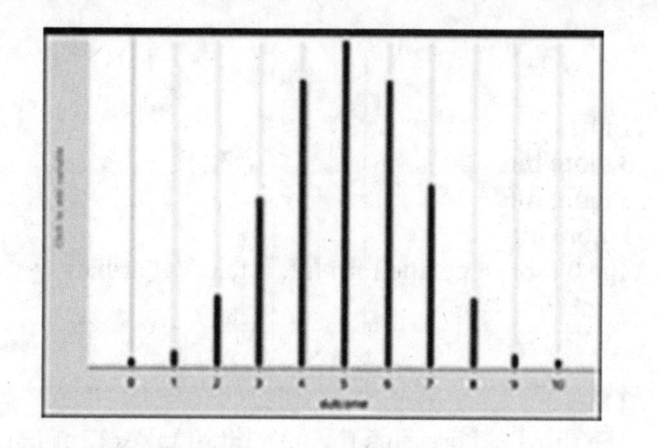

PROBLEM Mrs. Favata plays recreational softball and has an average of 0.400 this season. Create a simulation to estimate the number of at-bats she will need in order to get 10 more hits.

SOLUTION With her average being 0.400, we can designate the digits 0–3 as indicating she got a hit and 4–9 as she did not. We can generate sets of random numbers using our graphing calculator to get our results. For simplicity's sake, we will repeat the simulation 10 times and take an average of the number of at-bats needed to get 10 hits. In each case, the command used will be randint(0,9, 10). If you are using a TI-Nspire, first type RandSeed 12345 to seed the random number generator. If you are using a TI-84, type 12345 → rand to seed the random number generator. If you do this, you should get the same set of numbers that follows.

First Trial

9,0,0,6,1,3,3,4,4,1	There are a total of 6 values 0–3 so that is 6 hits.
4,9,2,1,6,2,6,3,**1**,4	She gets a hit with a 2, 1, 2, and 1. Counting the 10 at-bats from line 1 with the 9 at-bats until the second 1 is reached yields a total of 19 at-bats.

Second Trial

6,4,9,2,7,8,5,0,7,2	3 hits
0,7,1,5,4,7,8,2,7,3	4 more hits
6,9,9,7,6,2,1,5,4,8	2 more hits
0,1,5,2,7,9,4,7,4,8	The 0 represents the 10th hit. It takes 31 at-bats.

Third Trial

2,6,1,6,5,7,5,4,4,6	2 hits
4,9,2,1,9,3,3,6,5,7	4 more hits
2,6,2,8,5,5,9,9,0,**2**	The third 2 represents the 10th hit. It takes 30 at-bats.

Fourth Trial

2,2,8,4,4,1,8,1,2,0	6 hits
3,4,6,8,4,9,7,6,5,8	1 more hit
2,3,4,**3**,8,2,7,0,5,7	The second 3 represents the 10th hit. It takes 24 at-bats.

Fifth Trial

3,9,9,5,8,5,6,0,8,7	2 hits
6,5,6,4,2,1,5,2,8,7	3 more hits
9,8,6,2,6,3,5,3,8,7	3 more hits
2,6,6,5,6,7,7,9,4,6	1 more hit
0,2,7,5,9,4,1,5,2,5	The 0 represents the 10th hit. It takes 41 at-bats.

Sixth Trial

0,4,7,2,7,5,2,4,7,3	4 hits
0,2,3,2,3,8,**2**,9,3,3	The third 2 represents the 10th hit. It takes 17 at-bats.

Seventh Trial

6,0,3,7,5,7,8,5,2,5	3 hits
6,9,4,6,6,2,6,4,2,9	2 more hits
6,5,1,9,9,7,9,2,0,6	3 more hits
3,**1**,4,6,5,3,9,8,0,3	The 1 represents the 10th hit. It takes 32 at-bats.

Eighth Trial

1,7,7,4,2,1,6,3,9,3	5 hits
0,8,6,5,3,7,1,8,9,5	3 more hits
6,9,4,0,**0**,7,0,9,7,6	The second 0 represents the 10th hit. It takes 25 at-bats.

Ninth Trial

0,9,5,6,3,6,2,8,8,7	3 hits
4,7,8,4,5,5,0,0,1,9	3 more hits
8,3,9,5,2,5,0,4,7,**1**	The 1 represents the 10th hit. It takes 30 at-bats.

Tenth Trial

8,2,4,6,2,3,9,8,6,0	4 hits
1,6,4,9,6,0,2,9,5,8	3 more hits
6,0,6,6,0,**1**,4,1,3,0	The first 1 represents the 10th hit. It takes 26 at-bats.

According to this simulation, it will take Mrs. Favata an average of 27.5 at-bats to get 10 hits.

Of course, to get a better estimate, use a computer to repeat this process a large number of times.

The simulation could also be done without a computer. You could use a jar with a large number of marbles. For example, have a jar with 40 blue and 60 red marbles. Randomly select a marble from the jar. Continue to do so until 10 blue marbles are drawn. Record the total number of marbles drawn from the jar to represent the number of at-bats. Return the marbles to the jar. Shake the jar to redistribute the marbles before collecting the data for the next trial.

PROBLEM Maxie has a peanut butter and jelly sandwich for school lunch 15% of the time, and he does so on a random basis. Describe a simulation that can be used to determine the number of times Maxie has peanut butter and jelly for lunch during a month in which there are 20 school days.

SOLUTION Designate 0–14 as days that Maxie has peanut butter and jelly for lunch and 15–99 as days that he does not. Seed the random number generator to ensure randomness. A command such as randint(0,99,20) can represent one trial. Tally the number of times the numbers 0–14 appear in each trial.

EXERCISE 11·6

Describe a simulation that can be used to determine an answer to each of these problems.

1. Major Soggy cereal advertises that you get a plastic dinosaur with every box of cereal that you buy. There are five different dinosaurs in all. How many boxes of cereal would you expect to buy, on average, to get the complete set of dinosaurs?
2. The teams in the NBA Western division are considered better than the teams in the Eastern division. In a given year, the teams from the West beat the East in 55% of the games played. The final round of the NBA has a team from each division play each other in a best 4 out of 7 series. Determine the number of games that need to be played to determine a winner.

Trigonometry: Right triangles and radian measure

The study of similarity within right triangles provides the background for the study of trigonometry. The concept is extended when the triangles are put on the coordinate plane, and the triangles are located somewhere other than the first quadrant. A new measure for measuring angles, radian measure, is based strictly on the measurements of the geometric figures involved.

Right triangle trigonometry

Trigonometry started as an application of similar right triangles. The three basic functions—sine, cosine, and tangent—are remembered with the mnemonic **SOHCAHTOA**: The sine ratio (S) is the ratio of the lengths of the side opposite (O) to the acute angle to the hypotenuse (H); the cosine (C) is the ratio of the adjacent (A) side to the hypotenuse (H); and the tangent (T) is the ratio of the opposite (O) side to the adjacent (A) side. Each ratio is abbreviated with three letters: The sine ratio is sin, the cosine ratio is cos, and the tangent ratio is tan.

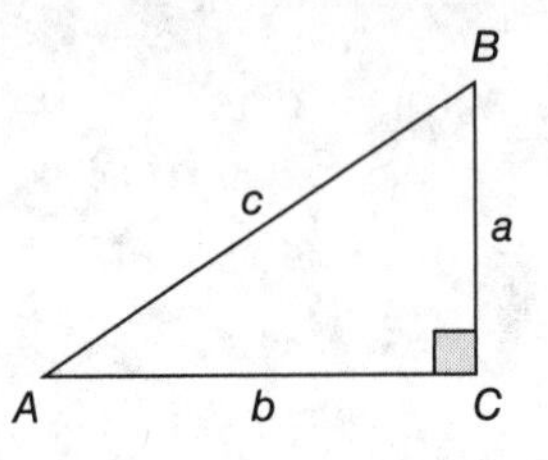

In the diagram shown, angle C is a right angle. The sine of angle A, written $\sin(A)$, is $\frac{a}{c}$; the cosine of angle A, $\cos(A)$, is $\frac{b}{c}$; and the tangent of angle A, $\tan(A)$, is $\frac{a}{b}$. Similarly, $\sin(B) = \frac{b}{c}$, $\cos(B) = \frac{a}{c}$, and $\tan(B) = \frac{b}{a}$.

Angles A and B are complementary angles. In fact, the word *cosine* comes from complement of the sine. For any acute angle, it will always be the case that $\sin(A) = \cos(90 - A)$.

PROBLEM Given the right triangle in the accompanying diagram, find $\overline{BC}$ to the nearest tenth.

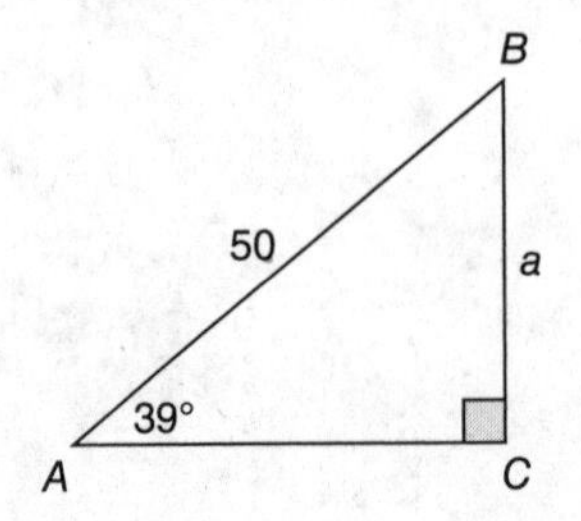

SOLUTION $\overline{BC}$ is the side opposite the given angle, and the hypotenuse is 50. Therefore, $\sin(39) = \frac{a}{50}$, or $a = 50\sin(39)$. With your calculator set in degree mode, compute $50\sin(39)$ to equal 31.5.

PROBLEM Given the right triangle in the accompanying diagram, find DE to the nearest tenth.

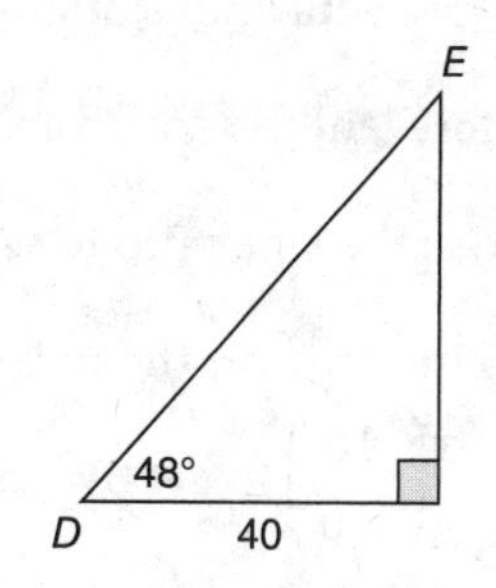

SOLUTION The length of the adjacent side is 40, and DE is the length of the hypotenuse.

Using the cosine ratio, write: $\cos(48) = \frac{40}{DE}$

Multiply: $DE \times \cos(48) = 40,$

$$DE = \frac{40}{\cos(48)}$$

Use your calculator to solve: $DE = 59.8.$

PROBLEM Find the length of $\overline{GH}$ in the accompanying diagram. Round your answer to the nearest tenth.

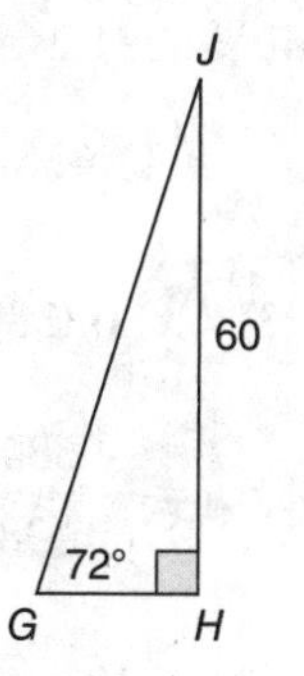

SOLUTION $JH = 60$ is opposite the angle and GH is adjacent to the angle.

Using the tangent ratio, write: $\tan(72) = \frac{60}{GH}$

Solve for GH: $GH = \frac{60}{\tan(72)} = 19.5$

You should notice that the measure of angle J is 18°. This problem could also be solved using the equation $\tan(18) = \frac{GH}{60}$, or $GH = 60\tan(18)$.

PROBLEM Find, to the nearest degree, the measure of θ in the accompanying diagram.

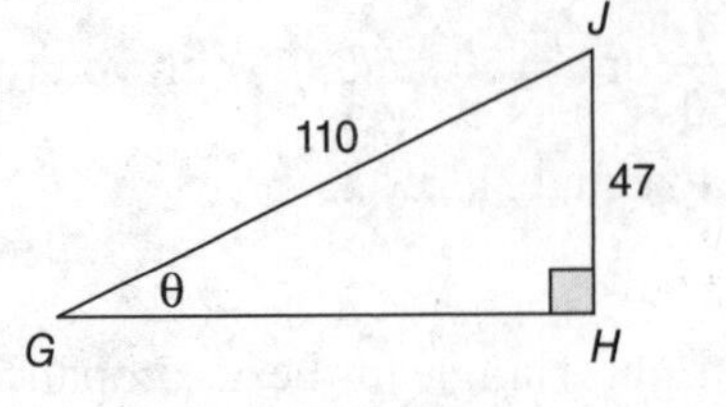

SOLUTION θ is the Greek letter **theta** and is often used in mathematics to represent angles.

The two sides whose lengths are given in the problem are the hypotenuse and the side opposite θ. Use the sine ratio to write $\sin(\theta)=\frac{47}{110}$. The inverse sine function, $\sin^{-1}$, on your calculator gives the measure of the acute angle whose sine is entered. Enter $\sin^{-1}\left(\frac{47}{110}\right)$ to get $\theta = 25°$.

PROBLEM The diagram represents the data collected to determine the height of a hill. The first angle measurement is taken at point A and is found to be 23°. $AB = 500$ ft, and a second angle measurement from point B to the top of the hill is 38°. Determine the height of the hill to the nearest foot.

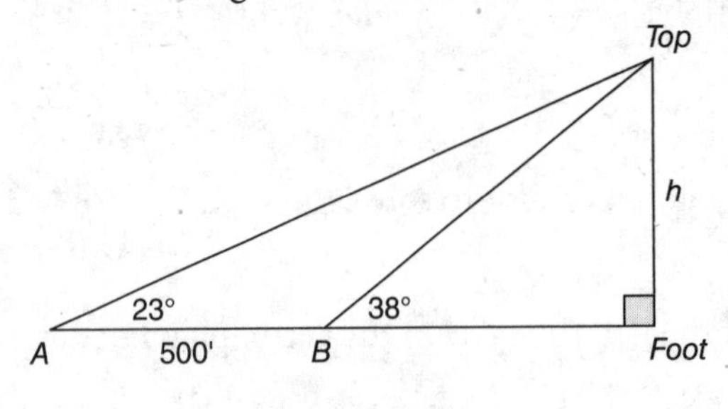

SOLUTION Let x represent the distance from B to the foot of the hill. $\tan(23) = \frac{h}{x+500}$, so that $(x + 500)\tan(23) = h$. $\tan(38) = \frac{h}{x}$, so $x\tan(38) = h$.

Solve $(x + 500)\tan(23) = x\tan(38)$: $x = \frac{500\tan(23)}{\tan(38) - \tan(23)}$. Substitute for x in the equation $h = x\tan 38)$: $h = \frac{500\tan(23)}{\tan(38) - \tan(23)}\tan(38) = 465$ ft.

EXERCISE 12·1

For questions 1–3, find the length of the missing side in each right triangle.

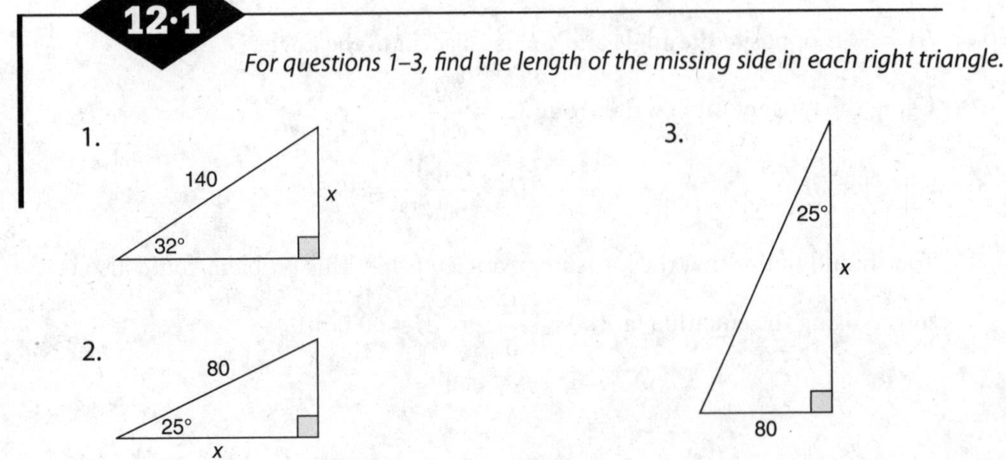

Find the measure of angle θ, to the nearest degree, for questions 4–6.

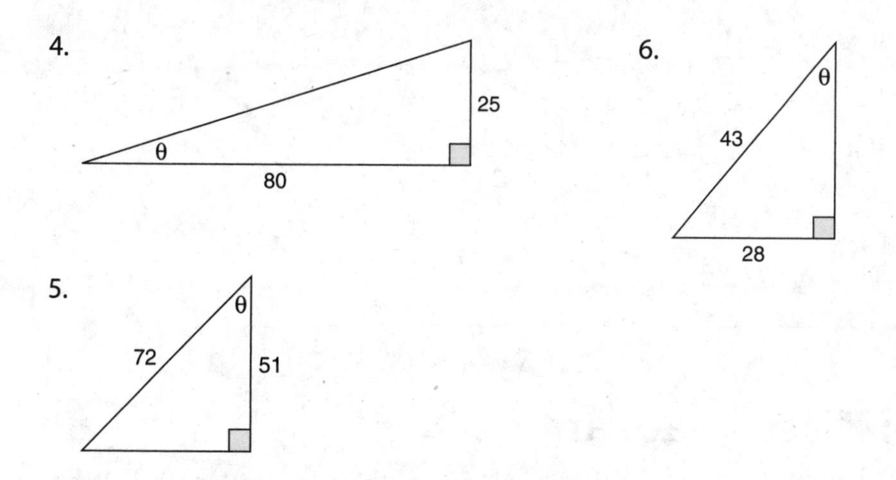

Meghan was hiking when she came to a flat region and saw a hill in the distance. Being a bit of a math lover, she happens to have an inclinometer in her backpack. She measures the angle of elevation to the top of the hill to be 18.3°. She walks directly toward the hill (in measured strides) another 400 ft. She measures the angle of elevation to the top of the hill to be 43.7°. She knows that she can now calculate the height of the hill when she gets back home.

7. Determine, to the nearest foot, the height of the hill.

Special right triangles

There are two triangles that have special importance in the study of trigonometry: the 30–60–90 and 45–45–90 triangles. The **30-60-90 triangle** is created by dividing an equilateral triangle into two congruent triangles, while the **45-45-90 triangle** (isosceles right triangle) is created by drawing in the diagonal of a square, which makes two congruent triangles.

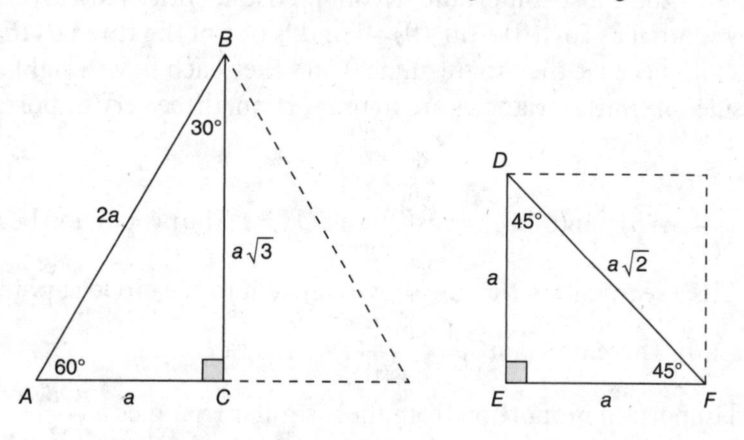

If each side of the equilateral triangle has length $2a$, then the length of the segment from A to midpoint C is a. Use the Pythagorean theorem to show that $BC = a\sqrt{3}$.

If each side of the square has length a, the length of the diagonal of the square is $a\sqrt{2}$. The values for the three trigonometric functions for the acute angles of these triangles are shown in the following table.

θ	sin(θ)	cos(θ)	tan(θ)
30°	$\frac{1}{2}$	$\frac{\sqrt{3}}{2}$	$\frac{1}{\sqrt{3}}=\frac{\sqrt{3}}{3}$
45°	$\frac{1}{\sqrt{2}}=\frac{\sqrt{2}}{2}$	$\frac{1}{\sqrt{2}}=\frac{\sqrt{2}}{2}$	1
60°	$\frac{\sqrt{3}}{2}$	$\frac{1}{2}$	$\sqrt{3}$

The unit circle: First quadrant

The **unit circle** is a circle with its center at the origin and with radius of length 1. When the right triangle is placed inside this circle, as shown in the figure, important relationships are determined.

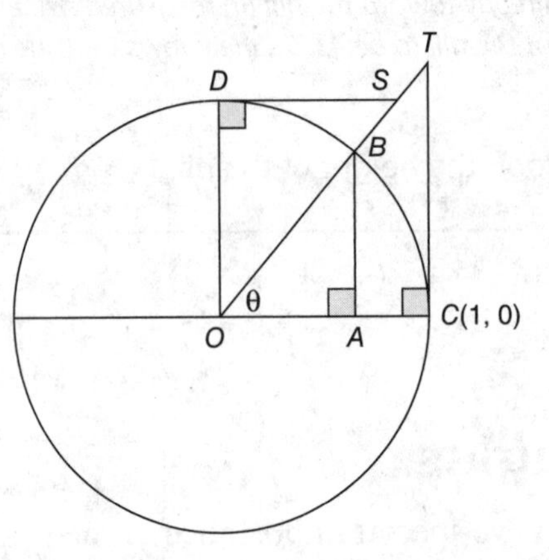

With the acute angle at the origin measuring θ, $OA = \cos(\theta)$, and $AB = \sin(\theta)$. The coordinates of point B are $(\cos(\theta), \sin(\theta))$. The equation of the unit circle is $x^2 + y^2 = 1$, and the coordinates for point B translate to the most important trigonometric identity $[\cos(\theta)]^2 + [\sin(\theta)]^2 = 1$. This is more commonly written as $\cos^2(\theta) + \sin^2(\theta) = 1$ and is one of the three **Pythagorean identities.**

$\Delta OAB \sim \Delta OCT$ because they share angle θ and they each have a right angle. Recalling that corresponding sides of similar triangles are in proportion, three very important relationships can be found.

First, $\frac{AB}{OA} = \frac{CT}{OC}$. $\overline{AB} = \sin(\theta)$, $OA = \cos(\theta)$, and $OC = 1$. The proportion becomes $\frac{\sin(\theta)}{\cos(\theta)} = \frac{CT}{1} =$ CT. However, CT is a segment on the line drawn tangent to the circle at point C. It is for this reason that $CT = \tan(\theta)$. Therefore, $\tan(\theta) = \frac{\sin(\theta)}{\cos(\theta)}$.

The second important proportion from these similar triangles is $\frac{OB}{OA} = \frac{OT}{OC}$. $OB = OC = 1$ and $OA = \cos(\theta)$. OT is a segment on the line which passes through the circle twice (extend the ray through O to get the second point of intersection). Such a line is called the **secant line**. Therefore, OT, which is the reciprocal of $\cos(\theta)$, is called the **secant of θ**, abbreviated $\sec(\theta)$. Therefore, $\sec(\theta) = \frac{1}{\cos(\theta)}$.

ΔOCT is a right triangle, so the Pythagorean theorem applies to it. $OC^2 + CT^2 = OT^2$. Substituting the new trigonometric values just learned, this equation becomes the second Pythagorean identity, $1+\tan^2(\theta)=\sec^2(\theta)$.

$\overline{SD} \parallel \overline{OC}$, because they are both perpendicular to the y-axis. Therefore, the measure of angle DSO must also be θ, as it is an alternate interior angle to angle BOA, and $\Delta OAB \sim \Delta SDO$. The ratio of corresponding sides yields two new trigonometric functions. $\frac{SD}{OS}=\frac{OA}{OB}$. With $OS = 1$ and SD being the line drawn tangent from the angle complementary to angle θ, $\overline{SD}$ is called the **cotangent function**, $\cot(\theta)$, so that $\cot(\theta)=\frac{\cos(\theta)}{\sin(\theta)}=\frac{1}{\tan(\theta)}$. By the same token, OS is called the **cosecant of θ**, $\csc(\theta)$, so that $\csc(\theta)=\frac{1}{\sin(\theta)}$. Finally, ΔSDO is a right triangle and $OD^2 + DS^2 = OS^2$, or $1+\cot^2(\theta)=\csc^2(\theta)$, which is the third Pythagorean identity.

PROBLEM Given acute angle A with $\sin(A)=\frac{3}{5}$, determine the values of the five remaining trigonometric functions.

SOLUTION Because $\sin(A)=\frac{3}{5}$ and $\csc(A)$ is the reciprocal of $\sin(A)$, $\csc(A)=\frac{5}{3}$. Use the Pythagorean identity $\cos^2(\theta)+\sin^2(\theta)=1$ to get $\cos^2(A)+\left(\frac{3}{5}\right)^2=1$. Solve this to get $\cos(A)=\frac{4}{5}$. The reciprocal of the cosine function is secant, so $\sec(A)=\frac{5}{4}$. Use $\tan(\theta)=\frac{\sin(\theta)}{\cos(\theta)}$ to get $\tan(A)=\frac{3}{4}$, and $\cot(\theta)=\frac{1}{\tan(\theta)}$ to get $\cot(A)=\frac{4}{3}$.

PROBLEM Given acute angle B with $\tan(B)=\frac{1}{2}$, determine the values of the remaining five trigonometric functions.

SOLUTION Cotangent is the reciprocal of tangent, so $\cot(B) = 2$. Use the Pythagorean identity $1+\tan^2(\theta)=\sec^2(\theta)$ to get $1+\left(\frac{1}{2}\right)^2=\sec^2(B)$, $\frac{5}{4}=\sec^2(B)$, so $\sec(B)=\frac{\sqrt{5}}{2}$. The reciprocal of secant is cosine, so $\cos(B)=\frac{2}{\sqrt{5}}=\frac{2\sqrt{5}}{5}$. $\tan(\theta)=\frac{\sin(\theta)}{\cos(\theta)}$ can be used to determine $\sin(B)$. $\sin(\theta)=\tan(\theta)\cos(\theta)=\left(\frac{1}{2}\right)\left(\frac{2}{\sqrt{5}}\right)=\frac{1}{\sqrt{5}}=\frac{\sqrt{5}}{5}$. Cosecant is the reciprocal of sine, so $\csc(B)=\sqrt{5}$.

EXERCISE 12·2

Complete the chart.

θ	csc(θ)	sec(θ)	cot(θ)
30°	1.	2.	3.
45°	4.	5.	6.
60°	7.	8.	9.

Given the acute angle Z with $\cos(Z) = \frac{7}{25}$*, find the value of the trigonometric functions specified in questions 10–14.*

10. $\sec(Z)$
11. $\sin(Z)$
12. $\csc(Z)$
13. $\tan(Z)$
14. $\cot(Z)$

Given the acute angle X with $\csc(X) = \frac{9}{8}$*, find the value of the trigonometric functions specified in questions 15–19.*

15. $\sin(X)$
16. $\cos(X)$
17. $\sec(X)$
18. $\tan(X)$
19. $\cot(X)$

The unit circle—beyond the first quadrant

Angles of rotation become an issue when angles move beyond the first quadrant. When drawn in **standard position**, the initial side of the angle is the positive x-axis. If the terminal side is drawn in a counterclockwise manner from the initial side, the angle is said to have positive measure; if drawn in a clockwise manner, the angle has negative measure. Angles whose terminal sides are the same ray are called **coterminal angles**. For example, an angle with measure 130° and an angle with measure –230° are coterminal. These same angles are coterminal with angles having measures 490°, 850°, 1210°, –590°, and –950°. It is possible to have angles with measure greater than 360° (think about a car spinning on ice—"doing a 360"—more than one revolution yields an angle that is more than 360°).

PROBLEM Find two angles, one with positive measure and one with negative measure, which are coterminal with an angle whose measure is 215°.

SOLUTION $360° + 215° = 575°$ will be an angle that is coterminal with 215°. $215° - 360° = -145°$ will also be an angle that is coterminal with 215°.

When the terminal side of θ goes beyond the first quadrant, the rules for opposite, adjacent, and hypotenuse need to be reconsidered. For example, when $\theta = 90°$, the coordinates for point B are (0, 1). Therefore, $\cos(90) = 0$ and $\sin(90) = 1$. $\csc(90)$ is also 1, while $\sec(90)$ and $\tan(90)$ are both undefined. (Stop a moment to think about this from an algebraic perspective, $\frac{1}{0}$ is undefined. From a geometric perspective, $\overline{OD} \parallel \overline{CT}$ so there are no points of intersection.) Angles that terminate on one of the axes are called **quadrantal angles.**

θ	$\sin(\theta)$	$\cos(\theta)$	$\tan(\theta)$	$\csc(\theta)$	$\sec(\theta)$	$\cot(\theta)$
0°	0	1	0	Undefined	1	Undefined
90°	1	0	Undefined	1	Undefined	0
180°	0	–1	0	Undefined	–1	Undefined
270°	–1	0	Undefined	–1	Undefined	0

If θ terminates within one of the quadrants, reflexive symmetry is used from a corresponding point in the first quadrant to determine the trigonometric values.

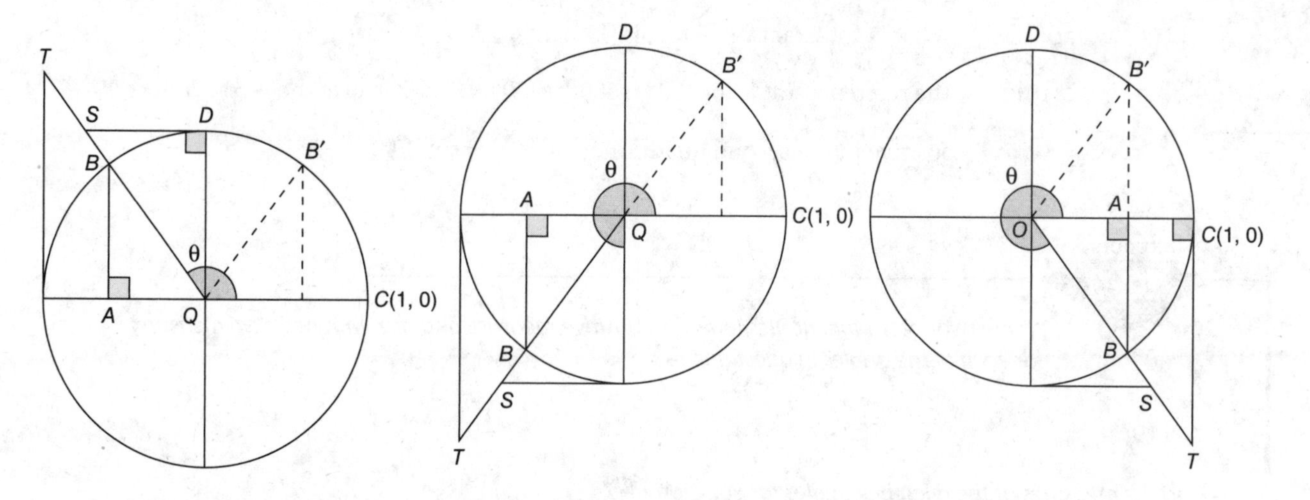

In each case, when point B is reflected back into the first quadrant, a triangle congruent to ΔOAB is formed. The acute angle in the first quadrant is called the **reference angle** for θ. It is imperative that you notice that the reference angle is always the acute angle that the terminal side of θ makes with the x-axis.

PROBLEM Find the reference angle for 125°.

SOLUTION An angle with measure 125° terminates in the second quadrant. The reference angle is formed with the negative x-axis (a 180° angle), so the measure of the reference angle is $180° - 125° = 55°$.

PROBLEM Find the reference angle for 219°.

SOLUTION An angle with measure 219° terminates in the third quadrant. The reference angle is formed with the negative x-axis (180°), so the reference angle is $219° - 180° = 39°$.

PROBLEM Find the reference angle for 310°.

SOLUTION An angle with measure 310° terminates in the fourth quadrant. The reference angle is formed with the positive x-axis (360°), so the reference angle is $360° - 310° = 50°$.

PROBLEM Express $\sin(219°)$ as a function of a positive acute angle.

SOLUTION The positive acute angle is always the reference angle. In this case, the acute angle will have measure 39°. In the third quadrant, the y-coordinate of $B[\sin(B)]$ is negative. Therefore, $\sin(219°) = -\sin(39°)$.

PROBLEM Express $\cos(310°)$ as a function of a positive acute angle.

SOLUTION Since cosine represents the x-coordinate of B, $\cos(310°)$ will be positive. With a reference angle of 50°, $\cos(310°) = \cos(50°)$.

PROBLEM Express $\tan(125°)$ as a function of a positive acute angle.

SOLUTION Tangent is the ratio of sine to cosine (or the ratio of the y-coordinate to the x-coordinate). In the second quadrant, this ratio will be negative. Therefore, $\tan(125°) = -\tan(55°)$.

PROBLEM What is the exact value for $\sin(150°)$?

SOLUTION The reference angle for 150° is $180° - 150° = 30°$. Because the y-coordinate is positive in the second quadrant, $\sin(150°) = \sin(30°) = \frac{1}{2}$.

EXERCISE 12·3

Name two coterminal angles (one with positive measure and one with negative measure) for each of the angles in questions 1–3.

1. 119°
2. 437°
3. −97°

Find the measures of the reference angles for questions 4–7.

4. 95°
5. 190°
6. 290°
7. 517°

For questions 8–10, express each trigonometric statement as a function of a positive acute angle.

8. $\tan(217°)$
9. $\sin(319°)$
10. $\cos(129°)$

For questions 11–13, determine the coordinates for each of the points on the unit circle.

11. 120°
12. 225°
13. 330°

For questions 14 and 15, find the specified trigonometric value.

14. Given $\sin(A) = \frac{-5}{13}$ with $\angle A$ terminating in quadrant III, find $\cos(A)$.
15. Given $\cos(B) = \frac{-9}{41}$ with $\angle B$ terminating in quadrant II, find $\tan(B)$.

Radian measure

By this point in your education, you probably have had many experiences dealing with different units of measurement. There is standard (inches, feet, yards, miles) versus metric (centimeter, meter, kilometer), as well as Fahrenheit versus Celsius. The use of the degree as a measure of angles goes back to the Babylonians and a calendar with 360 days. (A five-day religious celebration at the end of the year kept their calendar relatively accurate for their short time as a power in history.)

A more accurate measure of angle measure (despite its lack of usage among the general public) is the radian. A **central angle** is an angle whose vertex is the center of a circle and whose legs (sides) are radii intersecting the circle at two points. By definition, the **radian measure** of a central angle is the ratio of the arc formed by the angle to the length of the radius of that circle. A key piece of this definition is that the radian has no units. Arc length and radius will be measured in the same units (whether standard or metric) so will cancel each other within the ratio.

The conversion between degrees and radians is best considered when using a circle with radius 1 (although this is not necessary, as a dilation will create proportionally large radii and arcs without changing angles). A complete revolution of the circle is 360°, but it is also an arc of length 2π radians. Half a revolution is 180° or π radians. It is usually this reduced set of numbers that is used to convert from one measurement to the other: $\frac{\pi}{180} = \frac{\text{radian measure}}{\text{degree measure}}$.

PROBLEM Convert 60° to radians.

SOLUTION $\frac{\pi}{180} = \frac{r}{60}$ becomes $r = \frac{60\pi}{180} = \frac{\pi}{3}$.

Of course, this means that twice 60° will be twice $\frac{\pi}{3}$, so $120° = \frac{2\pi}{3}$, while half of 60° will be half $\frac{\pi}{3}$, or $30° = \frac{\pi}{6}$.

PROBLEM Convert $\frac{7\pi}{15}$ radians to degrees.

SOLUTION $\frac{\pi}{180} = \frac{\frac{7\pi}{15}}{d}$ becomes $\pi d = \left(\frac{7\pi}{15}\right)180$, so $\pi d = 84\pi$, and $d = 84°$.

If θ is the radian measure of the central angle, r is the length of the radius, and s is the length of the arc, the formula $\theta = \frac{s}{r}$ becomes $s = r\theta$.

PROBLEM A tire with a 16-inch radius makes 2000 revolutions. How far did the tire travel?

SOLUTION A point on the tire will travel a linear distance of $16(2\pi) = 32\pi$ inches after one revolution. After 2000 revolutions, the point on the tire will have traveled $64{,}000\pi$ in, which is 16,755 ft or 3.17 mi.

EXERCISE

12·4

Convert each of the angle measures in questions 1–3 to radian mode.

1. 72°
2. 140°
3. 315°

Convert each of the angle measures in questions 4–6 to degree mode.

4. $\frac{2\pi}{9}$
5. $\frac{5\pi}{18}$
6. $\frac{5\pi}{12}$

Given the information in questions 7 and 8, solve for the specified values.

7. In a circle with radius 8 cm, a central angle forms an arc with length 20 cm. Find the radian measure of the central angle.
8. A central angle with measure $\frac{3\pi}{5}$ forms an arc with length 4 cm. Find the length of the radius of the circle.

The Unit Circle – Let's put it all together

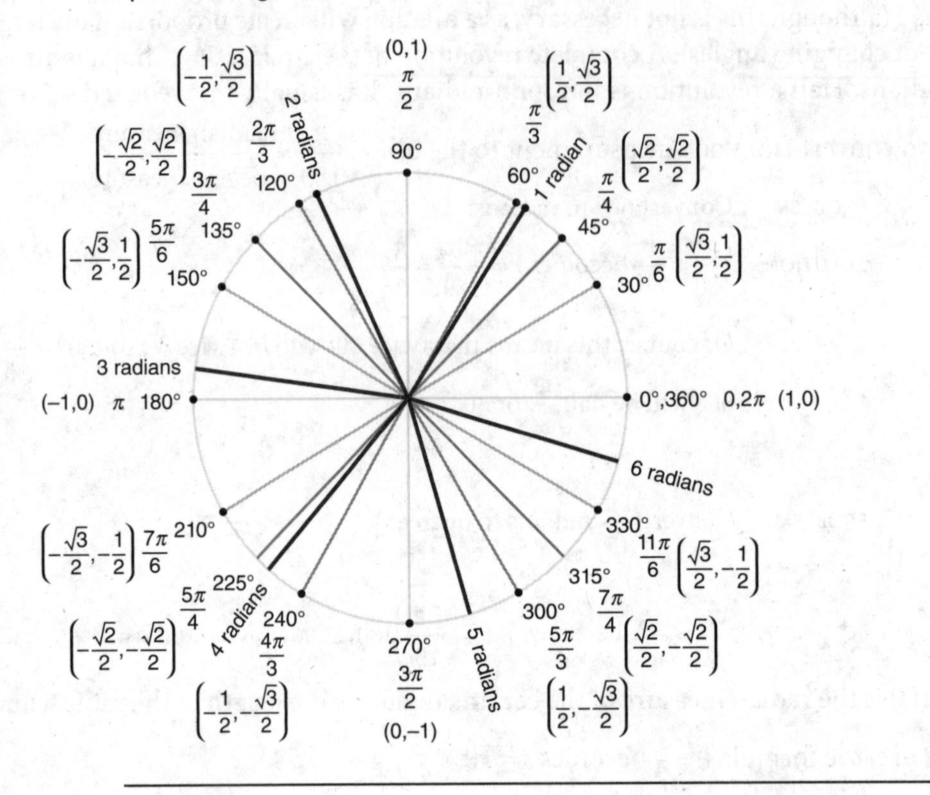

Basic trigonometric identities

Recapping the chapter so far, there are 22 relationships that have been established with regard to the trigonometric functions.

Quotient identities

$$\tan(\theta) = \frac{\sin(\theta)}{\cos(\theta)}$$

$$\cot(\theta) = \frac{\cos(\theta)}{\sin(\theta)}$$

Reciprocal identities

$$\sec(\theta) = \frac{1}{\cos(\theta)}$$

$$\csc(\theta) = \frac{1}{\sin(\theta)}$$

$$\cot(\theta) = \frac{1}{\tan(\theta)}$$

Pythagorean identities

$$\cos^2(\theta)+\sin^2(\theta)=1$$

$$1+\tan^2(\theta)=\sec^2(\theta)$$

$$1+\cot^2(\theta)=\csc^2(\theta)$$

Reflections

$$\sin(180^\circ-\theta)=\sin(\theta), \text{ or } \sin(\pi-\theta)=\sin(\theta)$$
$$\cos(180^\circ-\theta)=-\cos(\theta), \text{ or } \cos(\pi-\theta)=-\cos(\theta)$$

This can be extended for the other functions as well as other quadrants.

In addition to these relationships, there is the issue of counterclockwise versus clockwise rotations of the terminal side of the angle.

Rotational

$$\sin(-\theta)=-\sin(\theta)$$
$$\cos(-\theta)=\cos(\theta)$$
$$\tan(-\theta)=-\tan(\theta)$$

Look at the diagrams for angles terminating in the different quadrants to see this.

There are a number of other important formulas that you will need to be comfortable with applying.

Addition and subtraction formulas

$$\sin(A+B)=\sin(A)\cos(B)+\sin(B)\cos(A)$$
$$\sin(A-B)=\sin(A)\cos(B)-\sin(B)\cos(A)$$
$$\cos(A+B)=\cos(A)\cos(B)-\sin(A)\sin(B)$$
$$\cos(A-B)=\cos(A)\cos(B)+\sin(A)\sin(B)$$
$$\tan(A+B)=\frac{\sin(A+B)}{\cos(A+B)}=\frac{\tan(A)+\tan(B)}{1-\tan(A)\tan(B)}$$
$$\tan(A-B)=\frac{\sin(A-B)}{\cos(A-B)}=\frac{\tan(A)-\tan(B)}{1+\tan(A)\tan(B)}$$

Double angles

$$\sin(2A)=2\sin(A)\cos A)$$
$$\cos(2A)=\cos^2(A)-\sin^2(A)$$
$$=2\cos^2(A)-1$$
$$=1-2\sin^2(A)$$
$$\tan(2A)=\frac{\sin(2A)}{\cos(2A)}=\frac{2\tan(A)}{1-\tan^2(A)}$$

PROBLEM Given $\sin(A) = \frac{-40}{41}$ with $180° < A < 270°$, and $\cos(B) = \frac{3}{5}$, $270° < B < 360°$, find:

a. $\sin(A + B)$

b. $\cos(A + B)$

c. $\tan(A + B)$

SOLUTION With A in the third quadrant, $\cos(A) < 0$. Use the Pythagorean identity $\cos^2(\theta) + \sin^2(\theta) = 1$ to get $\cos^2(A) + \left(\frac{-40}{41}\right)^2 = 1$ to show that $\cos(A) = \frac{-9}{41}$. With angle B in the fourth quadrant, $\sin(B) < 0$. Use the same identity to show that $\sin(B) = \frac{-4}{5}$.

a. $\sin(A + B) = \sin(A)\cos(B) + \sin(B)\cos(A) = \left(\frac{-40}{41}\right)\left(\frac{3}{5}\right) + \left(\frac{-4}{5}\right)\left(\frac{-9}{41}\right) = \frac{-84}{205}$

b. $\cos(A + B) = \cos(A)\cos(B) - \sin(A)\sin(B) = \left(\frac{-9}{41}\right)\left(\frac{3}{5}\right) - \left(\frac{-40}{41}\right)\left(\frac{-4}{5}\right) = \frac{-187}{205}$

c. $\tan(A + B) = \frac{\sin(A+B)}{\cos(A+B)} = \frac{\frac{-84}{205}}{\frac{-187}{205}} = \frac{-84}{-187} = \frac{84}{187}$

PROBLEM Given $\cos(Z) = \frac{-8}{17}$ with $\frac{\pi}{2} < Z < \pi$, find:

a. $\sin(2Z)$

b. $\cos(2Z)$

c. $\tan(2Z)$

SOLUTION With Z in the second quadrant, use the Pythagorean identity to determine that $\sin(Z) = \frac{15}{17}$.

a. $\sin(2Z) = 2\sin(Z)\cos(Z) = 2\left(\frac{15}{17}\right)\left(\frac{-8}{17}\right) = \frac{-240}{289}$

b. $\cos(2Z) = \cos^2(Z) - \sin^2(Z) = \left(\frac{-8}{17}\right)^2 - \left(\frac{15}{17}\right)^2 = \frac{-161}{289}$

c. $\tan(2Z) = \frac{\sin(2Z)}{\cos(2Z)} = \frac{\frac{-240}{289}}{\frac{-161}{289}} = \frac{240}{161}$

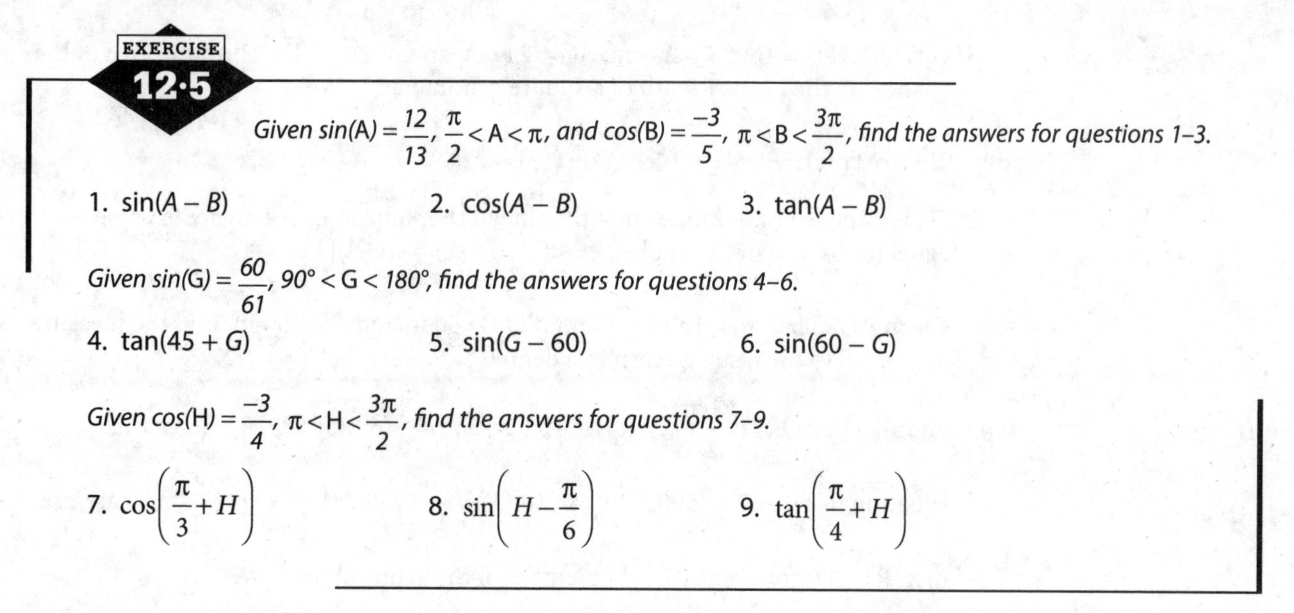

EXERCISE 12·5

Given $\sin(A) = \frac{12}{13}$, $\frac{\pi}{2} < A < \pi$, *and* $\cos(B) = \frac{-3}{5}$, $\pi < B < \frac{3\pi}{2}$, *find the answers for questions 1–3.*

1. $\sin(A - B)$
2. $\cos(A - B)$
3. $\tan(A - B)$

Given $\sin(G) = \frac{60}{61}$, $90° < G < 180°$, *find the answers for questions 4–6.*

4. $\tan(45 + G)$
5. $\sin(G - 60)$
6. $\sin(60 - G)$

Given $\cos(H) = \frac{-3}{4}$, $\pi < H < \frac{3\pi}{2}$, *find the answers for questions 7–9.*

7. $\cos\left(\frac{\pi}{3} + H\right)$
8. $\sin\left(H - \frac{\pi}{6}\right)$
9. $\tan\left(\frac{\pi}{4} + H\right)$

Area of a triangle

The area of ΔABC is $\frac{1}{2}(AB)h$. In right ΔADC, $\sin(A) = \frac{h}{AC}$ so that $h = AC\sin(A)$. Substitute this into the formula for the area of the triangle and get the result that the area of ΔABC is $\frac{1}{2}(AB)(AC)\sin(A)$. In general, the **area of a triangle** is equal to one-half the product of the lengths of two sides of a triangle and the sine of the included angle.

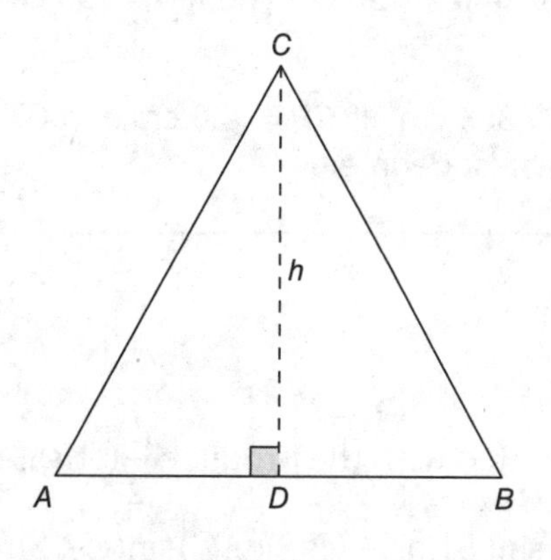

PROBLEM Given ΔABC with $AB = 20$ cm, $AC = 24$ cm, and $m\measuredangle A = 67°$, find the area of the triangle to the nearest tenth of a square centimeter.

SOLUTION The area of $\Delta ABC = \frac{1}{2}(20)(24)\sin(67) = 220.9 \text{ cm}^2$.

PROBLEM Given ΔABC with $AB = 20$ cm, $AC = 24$ cm, and $m\measuredangle A = 113°$, find the area of the triangle to the nearest tenth of a square centimeter.

SOLUTION The area of $\Delta ABC = \frac{1}{2}(20)(24)\sin(113) = 220.9 \text{ cm}^2$.

This example is a reminder that the sine of the supplement of an acute angle is equal to the sine of the angle [i.e., $\sin(180 - A) = \sin(A)$].

PROBLEM Given ΔQRS has an area of 560 in^2, if $QR = 60$ in and $RS = 50$ in, find the measure of angle R to the nearest tenth of a degree.

SOLUTION The area of ΔQRS is $\frac{1}{2}(\overline{QR})(\overline{RS})\sin(R)$, so $560 = \frac{1}{2}(60)(50)\sin(R)$. Solve this equation to get $560 = 1500\sin(R)$, or $\sin(R) = \frac{560}{1500}$. $\sin^{-1}\left(\frac{560}{1500}\right) = 21.9°$. Therefore, $m\measuredangle R = 21.9°$ or $158.1°$ (the supplement of the acute angle).

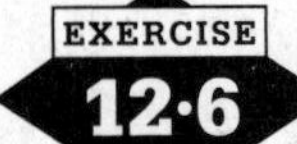

Using the information given, solve the following.

1. Given ΔABC, $AB = 20$ in, $AC = 30$ in, and $m\measuredangle A = 47°$, find the area of ΔABC to the nearest square inch.
2. Given ΔWHY with $\overline{WH} = 80$ ft, $\overline{HY} = 90$ ft, and $m\measuredangle H = 123°$, find the area of ΔWHY to the nearest square foot.
3. The area of ΔKLM is 8693 cm^2. If $\overline{KL} = 120$ cm and $\overline{KM} = 150$ cm, find the measure of angle K to the nearest degree.
4. The area of obtuse ΔGHJ is 71,365 cm^2. If $\overline{GH} = 420$ cm and $\overline{GJ} = 540$ cm, find the measure of the obtuse angle G to the nearest degree.

Law of sines

The area of ΔABC can be computed with the product: $\frac{1}{2}(AB)(AC)\sin(A)$. It can also be computed with the product $\frac{1}{2}(AB)(BC)\sin(B)$ or $\frac{1}{2}(BC)(AC)\sin(C)$. Since all three expressions represent the area of the same triangle, they must be equal.

$$\frac{1}{2}(AB)(AC)\sin(A) = \frac{1}{2}(AB)(BC)\sin(B) = \frac{1}{2}(BC)(AC)\sin(C).$$

Divide all three expressions by $\frac{1}{2}(AB)(BC)(AC)$ to get the equation that is called the **law of sines:** $\frac{\sin(A)}{BC} = \frac{\sin(B)}{AC} = \frac{\sin(C)}{AB}$. Using the convention of naming the side of a triangle with the lower-case letter matching the vertex of the angle, this becomes the more familiar form of the equation: $\frac{\sin(A)}{a} = \frac{\sin(B)}{b} = \frac{\sin(C)}{c}$.

It is important to note that the proportion created by any two fractions in the law of sines involves two sides and two angles, and that the orientation of these sides and angles fits the patterns AAS, ASA, or SSA, which you may remember from your study of geometry.

PROBLEM In ΔGEM, $\overline{EM} = 95$ cm, $m\angle G = 56°$, and $m\angle E = 49°$. Find the lengths of $\overline{GE}$ and $\overline{GM}$ to the nearest tenth of a centimeter.

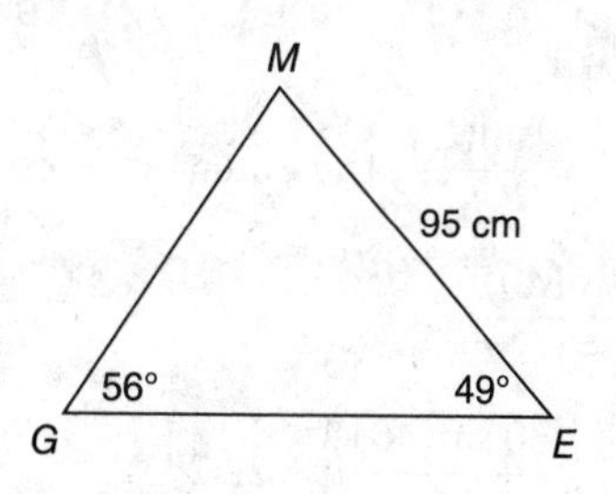

SOLUTION Subtract the sum of 56° and 49° from 180° to determine the measure of the third angle of the triangle, $m\angle M = 75°$.

Use the law of sines to solve for each side: $\dfrac{\sin(56)}{95} = \dfrac{\sin(49)}{GM}$

Multiply: $GM\sin(56) = 95\sin(49)$

$$GM = \frac{95\sin(49)}{\sin(56)} = 86.4 \text{ cm}$$

$\dfrac{\sin(56)}{95} = \dfrac{\sin(75)}{GE}$ becomes $GE\sin(56) = 95\sin(75)$ or $GE = \dfrac{95\sin(75)}{\sin(56)} = 110.7$ cm.

PROBLEM In ΔJET, $JE = 150$ in, $m\angle J = 67°$, and $m\angle E = 49°$. Find the lengths of $\overline{JT}$ and $\overline{ET}$ to the nearest tenth of an inch.

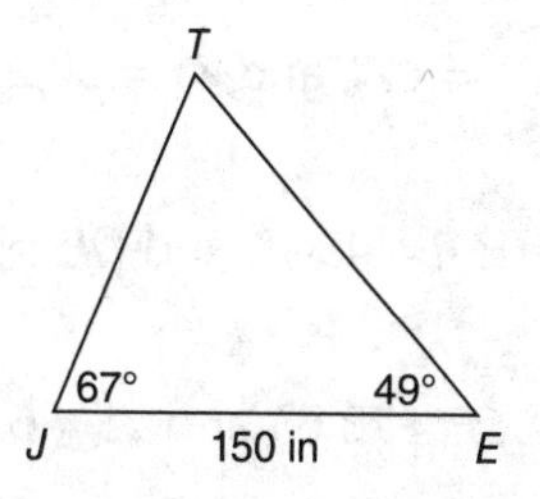

SOLUTION Subtract the sum of 67° and 49° from 180° to determine the measure of the third angle of the triangle, $m\angle T = 64°$.

Use the law of sines to solve for each side: $\dfrac{\sin(64)}{150} = \dfrac{\sin(67)}{ET}$

Multiply: $ET\sin(64) = 150\sin(67)$

$$ET = \frac{150\sin(67)}{\sin(64)} = 153.6 \text{ in}$$

$\dfrac{\sin(64)}{150} = \dfrac{\sin(49)}{JT}$ becomes $JT\sin(64) = 150\sin(49)$, so $JT = \dfrac{150\sin(49)}{\sin(64)} = 126.0$ in.

PROBLEM In ΔABC, $AB = 50$ cm, $BC = 30$ cm, and $m\measuredangle C = 40°$. Find, to the nearest tenth of a degree, the measure of $\measuredangle A$.

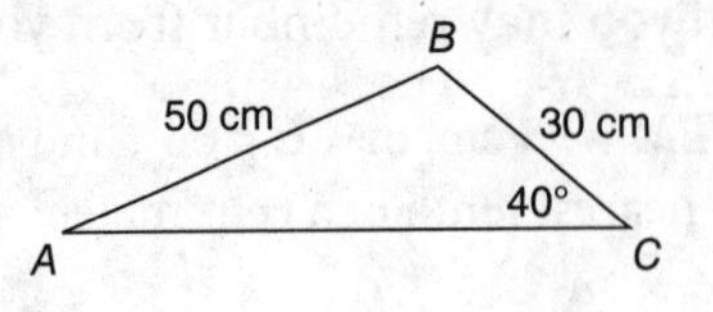

SOLUTION Using $\dfrac{\sin(A)}{a} = \dfrac{\sin(C)}{b}$, the equation becomes $\dfrac{\sin(A)}{30} = \dfrac{\sin(40)}{50}$ or $\sin(A) = \dfrac{30\sin(40)}{50}$.

$$m\measuredangle A = \sin^{-1}\left(\frac{30\sin(40)}{50}\right) = 22.7°$$

Because $AB > BC$, it must be the case that $m\measuredangle C > m\measuredangle A$ (this is one of the key concepts in geometry—the larger side of a triangle is always opposite the larger angle). Therefore, the only possible measure for $\measuredangle A$ is 22.7°.

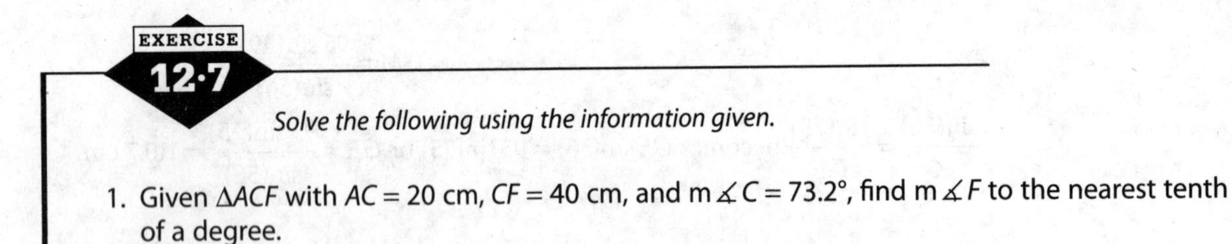

Solve the following using the information given.

1. Given ΔACF with $AC = 20$ cm, $CF = 40$ cm, and $m\measuredangle C = 73.2°$, find $m\measuredangle F$ to the nearest tenth of a degree.
2. Given ΔKLM with $KL = 90.2$ ft, $LM = 131.4$ ft, and $m\measuredangle K = 59°$, find $m\measuredangle M$ to the nearest tenth of a degree.
3. Given ΔXYZ with $m\measuredangle X = 67°$, $m\measuredangle Y = 47°$, and $XY = 50.2$ cm, find XZ and YZ to the nearest tenth of a centimeter.
4. Given ΔQRS with $m\measuredangle Q = 37.2°$, $m\measuredangle R = 43.1°$, and $QR = 11.7$ cm, find QS and RS to the nearest tenth of a centimeter.
5. Given ΔYES with $m\measuredangle Y = 47.1°$, $m\measuredangle E = 73.6°$, and $YS = 61.7$ cm, find ES and EY to the nearest tenth of a centimeter.
6. Given ΔQED with $m\measuredangle Q = 71.2°$, $m\measuredangle E = 63.1°$, and $QD = 53.2$ cm, find QE and ED to the nearest tenth of a centimeter.

Ambiguous case

Given $\measuredangle C$ and $BC = 30$ cm with $m\measuredangle C = 40°$, what is the length of the shortest segment that can be drawn from point B to create a triangle?

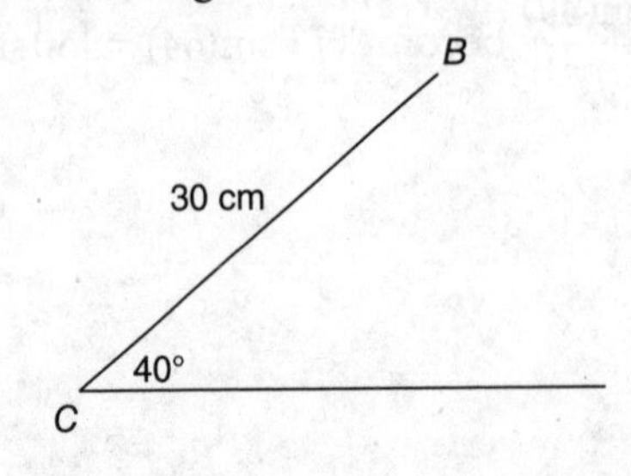

If arcs are drawn from vertex B, as shown in the accompanying diagram, there is one arc that intersects the ray of angle C only once.

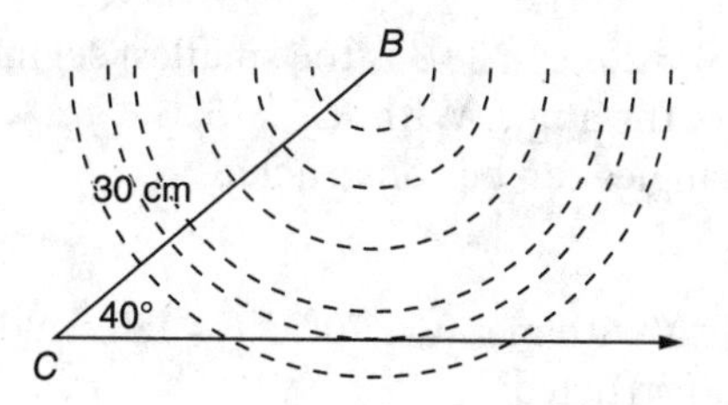

The point of intersection for this arc is the foot of the perpendicular from B to the ray.

> Hopefully, this makes sense to you because the shortest distance from a point to a line is along the perpendicular. This is why you are told to stand up "straight" when someone is measuring your height.

If point A is the point at which the perpendicular intersects the side of the ray, then ΔABC is a right triangle and $BA = 30\sin(40)$.

If an arc is drawn with a length AB that is less than the length of the perpendicular, $BC\sin(C)$, the arc will never intersect the side of the ray. It would not be possible to create a triangle with the given measurements.

If an arc is drawn with a length AB that is greater than BC, the arc will intersect the ray somewhere to the right and one triangle can be constructed.

If an arc is drawn with a length AB that is greater than $BC\sin(C)$ but less than BC [i.e., $BC\sin(C) < AB < BC$], then there will be two triangles that can be constructed. It is this scenario that is referred to as the **ambiguous case.**

The situations described above apply when the angle in question is an acute angle. If the given angle is either a right angle or an obtuse angle, the length of the arc drawn from the end-point of the segment must be longer than the segment itself or the arc will not intersect the other side of angle C.

PROBLEM For angle A with measure 45°, $AR = 40$ in, $RT = 20\sqrt{2}$ in. How many triangles, if any, can be constructed?

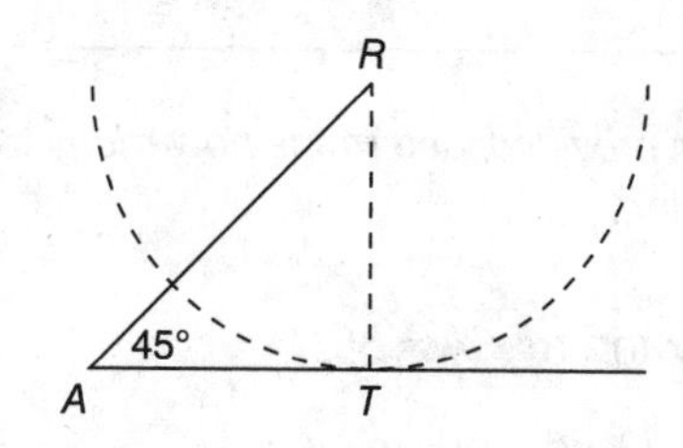

SOLUTION The shortest segment that can be drawn from R to make the triangle is along the perpendicular. The length of the perpendicular is $40\sin(45) = 20\sqrt{2}$. Therefore, you know that ΔART is a right triangle. There is one triangle that can be constructed.

PROBLEM For angle A with measure 45°, $AR = 40$ in, $RT = 25$ in. How many triangles, if any, can be constructed?

SOLUTION $40\sin(45) = 20\sqrt{2} \approx 28.3$ is the smallest segment that can be constructed to reach the side of the angle. With $RT = 25$, the arc will not intersect the side of the angle, so no triangles can be constructed.

PROBLEM For angle X with measure 70°, $XF = 150$ cm, $FM = 165$ cm. How many triangles can be constructed?

SOLUTION The arc drawn from point F is longer than XF, so one triangle can be constructed.

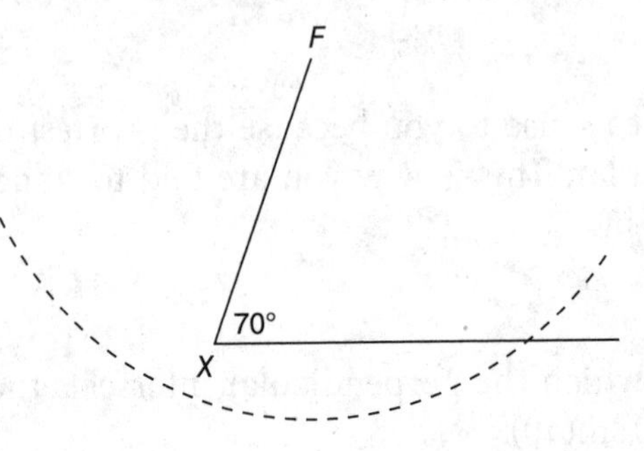

PROBLEM For angle X with measure 70°, $XF = 150$ cm, $FM = 145$ cm. How many triangles can be constructed?

SOLUTION The shortest segment from F that can form a triangle has length $150\sin(70) \approx 140.95$ cm. Because the length of FM is between $150\sin(70)$ and 150 (the length of XF), there are two triangles that can be constructed.

PROBLEM For angle P with measure 110°, $MP = 250$ m, $MG = 245$ cm. How many triangles can be constructed?

SOLUTION Since $\measuredangle P$ is an obtuse angle, MG must be longer than MP to create a triangle. Because $MG < MP$, no triangles can be constructed.

EXERCISE 12·8

Using the information provided, determine how many triangles, if any, can be constructed for each of the following.

1. In ΔKAT, $KA = 68$ in, $KT = 34$ in, and $m\measuredangle A = 30°$.
2. In ΔDAY, $DA = 126$ yd, $YD = 63\sqrt{3}$ yd, and $m\measuredangle A = 60°$.
3. In ΔTAD, $TA = 82$ mm, $TD = 44$ mm, and $m\measuredangle D = 40°$.
4. In ΔNET, $TE = 32$ cm, $TN = 44$ cm, and $m\measuredangle E = 130°$.
5. In ΔARG, $RA = 95$ m, $RG = 67$ m, and $m\measuredangle A = 110°$.
6. In ΔMBA, $MA = 72$ cm, $AB = 60$ cm, and $m\measuredangle M = 50°$.
7. In ΔEDU, $EU = 125$ mm, $DU = 114$ mm, and $m\measuredangle E = 57°$.

Law of cosines

The law of sines is used when the information available for the triangle fits one of the patterns ASA, AAS, or SSA. When the information available for the triangle fits the pattern SSS or SAS, the **law of cosines** is used to find the missing information about the triangle. The formula for the law of cosines is

$$p^2 = q^2 + r^2 - 2\,qr\cos(P)$$

The key to this formula is to realize that when the information fits the SAS pattern, it is the third side of the triangle that is determined first. When the information is of the SSS pattern, it is an angle that is determined first. In the formula for the law of cosines, the side and the angle opposite that side are at the beginning and end of the formula.

PROBLEM In ΔDUE, $DE = 90$ cm, $UE = 70$ cm, and $m\measuredangle E = 57°$. Find the length of $\overline{DU}$, to the nearest tenth of a centimeter.

SOLUTION $\overline{DU}$ is also known as side e. The information given for ΔDUE fits the SAS pattern, so use the law of cosines to get $e^2 = 90^2 + 70^2 - 2(90)(70)\cos(57)$. Use your calculator to compute the value for e^2 (approximately 6137.55). Take the square root of the value on the calculator screen to determine $e = DU = 78.3$ cm.

PROBLEM Given ΔSPY with $SP = 23.7$ m, $PY = 13.1$ m, and $YS = 19.2$ m, find $m\measuredangle Y$ to the nearest tenth of a degree.

SOLUTION The information provided for ΔSPY fits the SSS pattern.

Use the law of cosines and note that $\overline{SP}$ is opposite $\measuredangle Y$ in this triangle:

$23.7^2 = 13.1^2 + 19.2^2 - 2(13.1)(19.2)\cos(Y)$

$561.69 = 171.61 + 368.64 - 503.04\cos(Y)$

$561.69 = 540.25 - 503.04\cos(Y)$

The temptation is to subtract 503.04 from 540.25, but this is not mathematically correct. In the same way that one cannot simplify $5 - 3x$ because the terms are not "like" terms, 540.25 and $503.04\cos(Y)$ are not like terms.

Subtract 540.25 from both sides of the equation:

$21.44 = -503.04\cos(Y)$

Solve for $\cos(Y)$:

$\cos(Y) = \dfrac{21.44}{-503.44}$, so $Y = \cos^{-1}\left(\dfrac{21.44}{-503.44}\right) = 92.4°$.

The law of sines or the law of cosines can be used to determine $m\measuredangle S$ and $m\measuredangle P$ if you are directed to do so, although the law of sines would be less work computationally.

The law of sines and the law of cosines are used in the resolution of vectors. **Vectors** are represented as directed line segments (they have length and direction), and the sum of two vectors is accomplished using the parallelogram method. That is, the vectors are placed together at their "tails" (endpoints) with the direction of the vector indicated with an arrow as shown in the accompanying diagram.

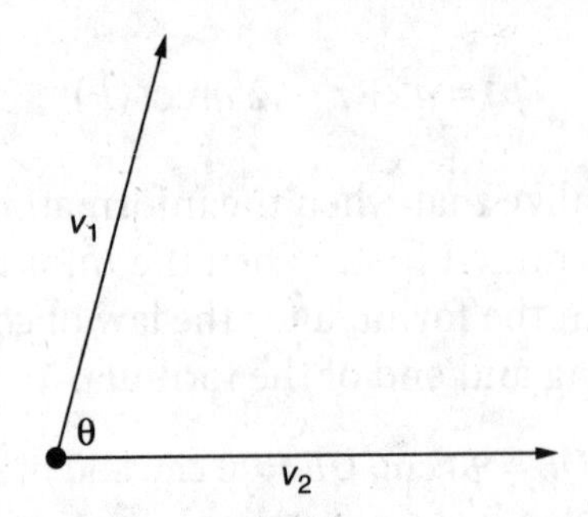

A parallelogram is constructed from these two sides, and the diagonal of the parallelogram drawn from the original set of tails is called the **resultant**, or net impact of the two vectors. The resulting picture is the diagram for a vector sum.

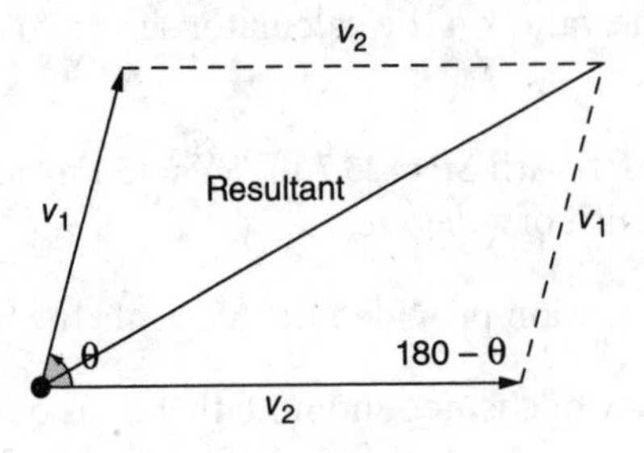

PROBLEM Two forces with magnitudes 60 N and 90 N act on an object with an angle of 70° between the two forces. Determine the magnitude of the resultant force, correct to the nearest tenth of a newton.

SOLUTION The first step to solving this problem is to construct the vector diagram as shown in the accompanying image. ΔABC displays SAS information. Use the law of cosines to determine the magnitude of the resultant, r.

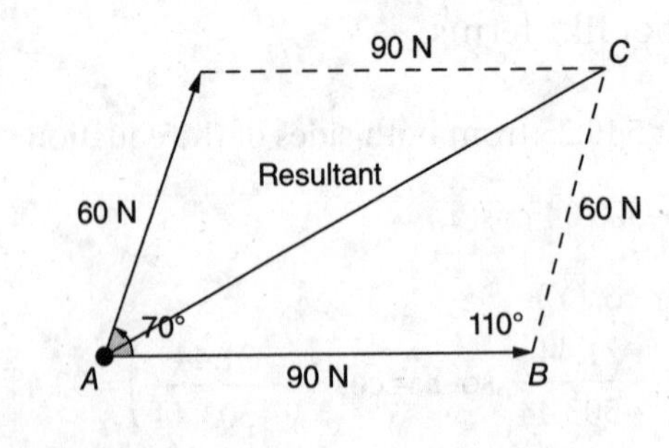

$$r^2 = 60^2 + 90^2 - 2(60)(90)\cos(110)$$
$$r^2 \approx 8006.18$$
$$r = 89.5 \text{ N}$$

PROBLEM Two forces of 140 lb and 190 lb act on an object with a resultant force of 150 lb. Determine, to the nearest tenth of a degree, the angle between the two forces.

SOLUTION Again, the first step to solving this kind of a problem is to draw the vector diagram as shown in the accompanying image. ΔHGI displays SSS information. Use the law of cosines to determine the measure of angle θ.

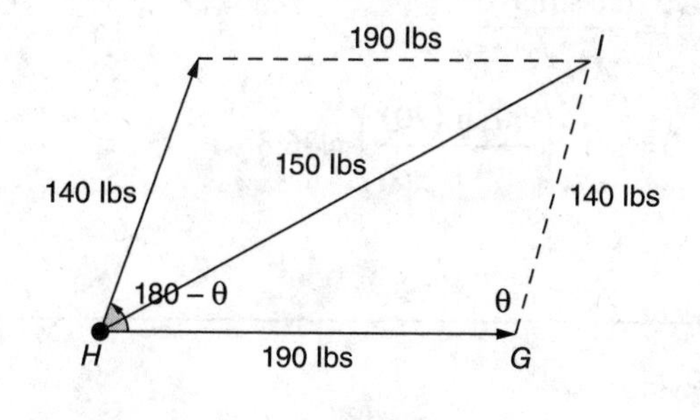

$$
\begin{aligned}
150^2 &= 140^2 + 190^2 - 2(140)(190)\cos(\theta) \\
22{,}500 &= 19{,}600 + 36{,}100 - 53{,}200\cos(\theta) \\
22{,}500 &= 55{,}700 - 53{,}200\cos(\theta) \\
-33{,}200 &= -53{,}200\cos(\theta) \\
\cos(\theta) &= \frac{33{,}200}{53{,}200} \\
\theta &= \cos^{-1}\left(\frac{33{,}200}{53{,}200}\right) = 51.39^\circ
\end{aligned}
$$

The angle between the two forces is the supplement of this angle, $180^\circ - 51.39^\circ = 128.6^\circ$.

PROBLEM Two forces of 50 N and 80 N act on an object with an angle of 110° between them. Find, to the nearest tenth of a degree, the measure of the angle formed between the resultant and the larger force.

SOLUTION Draw the vector diagram as shown in the accompanying image. ΔKJL displays SAS information. Use the law of cosines to determine the magnitude of the resultant, r.

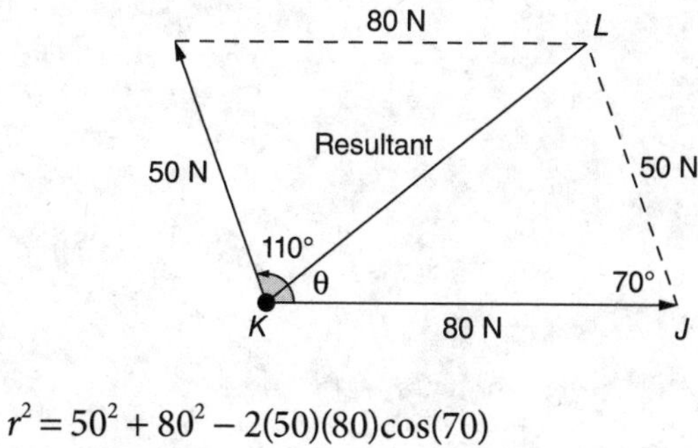

$$
\begin{aligned}
r^2 &= 50^2 + 80^2 - 2(50)(80)\cos(70) \\
r^2 &\approx 6163.84 \\
r &\approx 78.51 \text{ N}
\end{aligned}
$$

Because this is an intermediate step in the problem, you will use the full number of decimal places from your calculator in the next step.

The law of sines can be used to compute the value of θ.

$$\frac{r}{\sin(70)} = \frac{50}{\sin(\theta)}$$

$$r\sin(\theta) = 50\sin(70)$$

$$\sin(\theta) = \frac{50\sin(70)}{r}$$

$$\theta = \sin^{-1}\left(\frac{50\sin(70)}{r}\right) = 36.8°$$

EXERCISE

12·9

Solve the following.

1. Given ΔNEC with $NC = 20$ cm, $NE = 40$ cm, and $m\measuredangle N = 73.2°$, find EC to the nearest centimeter.
2. Given ΔMVP with $MV = 90.2$ ft, $MP = 131.4$ ft, and $m\measuredangle M = 59°$, find VP to the nearest foot.
3. Given ΔXYZ with $XY = 67$ mm, $YZ = 47$ mm, and $XZ = 50$ mm, find $m\measuredangle X$ to the nearest tenth of a degree.
4. Given ΔPTS with $PT = 37.2$ mm, $PS = 43.1$ mm, and $TS = 11.7$ mm, find $m\measuredangle T$ to the nearest tenth of a degree.
5. Vectors with magnitudes 83 N and 75 N act on an object at an angle of 40° to each other. Find the magnitude of the resultant force to the nearest tenth of a newton.
6. Vectors with magnitudes 100 N and 120 N act on an object with a resultant force of 160 N. Find, to the nearest degree, the angle between the two original forces.
7. Vectors with magnitudes 50 N and 70 N act on an object with a resultant force of 85 N. Find, to the nearest tenth of a degree, the angle between the resultant and the larger force.

Graphs of trigonometric functions

There are many examples of periodic phenomena in the world—ocean tides, alternating current, and the height of the rider off the ground in a Ferris wheel, to name a few. The graphs of these periodic phenomena usually require trigonometric functions.

Amplitude and period

Periodic phenomena are things you deal with every day. What time is it now? What will the time be 24 hours from now? One week from now? A month from now? A year from now? Phases of the moon (and therefore, the tides) and the location of the stars in the night sky are probably what first got the early civilizations to study periodic phenomena, as their livelihoods and lives depended on this knowledge. EKGs show the beating of the heart, and in a healthy heart, will show periodic tendencies.

Most studies of the periodic phenomena are tied to the trigonometric functions but are not limited to them. For example, square waves are important in the study of harmonics. Examine the square wave, $f(x)$, in the accompanying diagram.

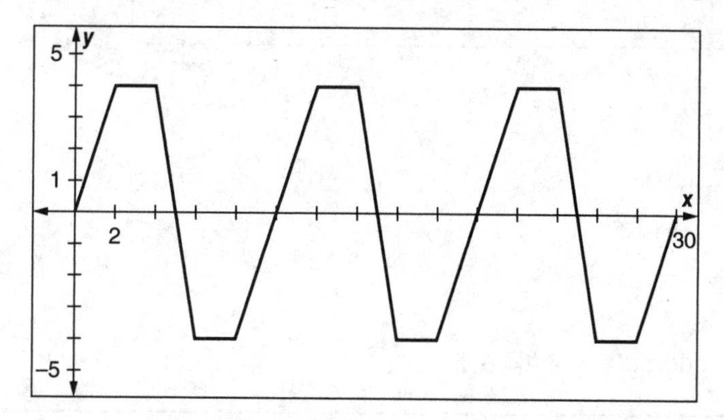

The graph repeats itself every 10 units. It is said that the **period** of this graph is 10. Knowing that $f(2) = 4$, it is then known that $f(12)$, $f(22)$, $f(222)$, $f(-8)$, and $f(-108)$ will also be 4. The amplitude of this function is defined as one-half the difference between the maximum and minimum values. For this function, the amplitude is $\frac{4-(-4)}{2} = 4$. The amplitude can also be thought of as the distance from the central value (average) to an extreme value; in this case, the maximum is 4 and the minimum is -4. Do you see that the graph of $g(x) = f(x) + 1$ will also have a period of 10 and an amplitude of 4, even though the average value will now be 1? Translating the graph vertically will change the average value, but it will not change the amplitude or period.

PROBLEM What are the amplitude, period, and average value of the function $k(x) = 3f(x) - 2$, with $f(x)$ being the function defined in the previous graph?

SOLUTION The graph of $k(x)$ is created by stretching the graph of $f(x)$ by a factor of 3 from the x-axis, and then translating it down 2 units. This transformation has no impact on the input values, so the period remains as 10. The range of $f(x)$ is $-4 \le y \le 4$, so the range for $k(x)$ is $3(-4) - 2 \le y \le 3(4) - 2$ or $-14 \le y \le 10$. The amplitude is $\frac{10-(-14)}{2} = 12$. The amplitude of the graph of $k(x)$ is 3 times the amplitude for the graph of $f(x)$. The average value of the graph is -2, the amount by which the graph was translated vertically.

EXERCISE 13·1

Use the accompanying graph of the function f(x) to determine the answers to questions 1–5.

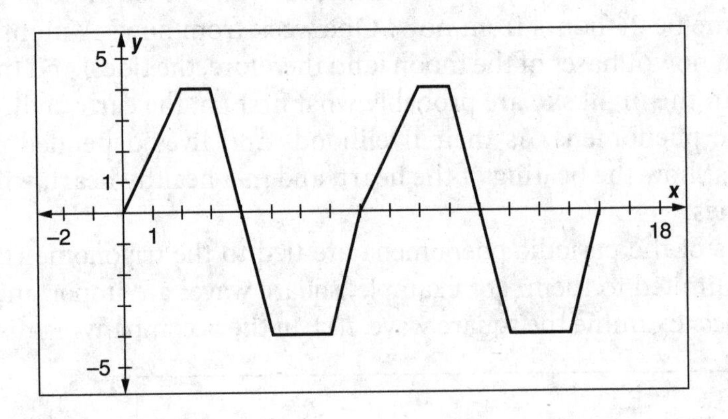

1. The period for the graph
2. The amplitude of the graph
3. The value of $f(19)$
4. The value of $f(-203)$
5. The amplitude of the function $g(x) = 2f(x) - 1$

Graphing trigonometric functions

The graphs of the three basic functions $y = \sin(x)$, $y = \cos(x)$, and $y = \tan(x)$ are shown in the following images. Each function is graphed on the interval $-2\pi \leq x \leq 2\pi$.

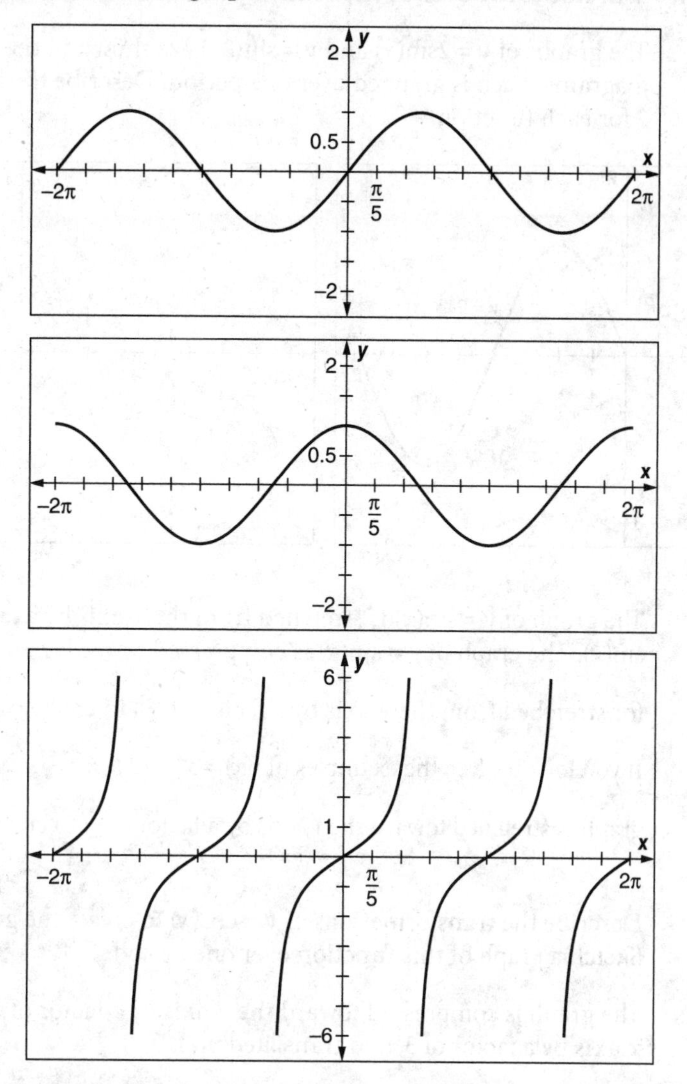

The period for the sine and cosine functions is 2π, while the period for the tangent function is π.

The amplitude is 1 for the sine and cosine functions. The amplitude is infinitely large for the tangent function—it is commonly said that the tangent function does not have an amplitude because it is not finite.

Transformations of the sine and cosine graphs are written in the form $y = A\sin(Bx + C) + D$ and $y = A\cos(Bx + C) + D$. The amplitude for each function is $|A|$, the period is $\frac{2\pi}{|B|}$, the vertical translation is D, and the horizontal translation, called the **phase shift**, is $\frac{-C}{B}$.

You probably have had plenty of experience working with the graphs of the form $y = af(x)$. The impact of a is to stretch the graph from the x-axis by a factor of a. The graphs of equations of the form $y = f(bx)$ most likely have been less frequent in your experience. For example, let $f(x) = x^2$, $a = 3$, and $b = 2$. The graph of $y = af(x) = 3x^2$ is the parabola stretched from the x-axis by a factor of 3. The graph of $y = f(bx) = (2x)^2 = 4x^2$ is the parabola stretched from the x-axis by a factor of 4.

If $f(x) = \sqrt{x}$, then the graph of $f(bx)$ would be the graph of the square root function stretched from the x-axis by a factor of $\sqrt{2}$. While there is nothing wrong with these explanations for the algebraic functions, they will not carry over to the study of the trigonometric functions.

PROBLEM The graphs of $y = 2\sin(x)$ and $y = \sin(2x)$ are shown in the accompanying diagrams. Each is graphed over one period. Describe the impact of the coefficient 2 for each function.

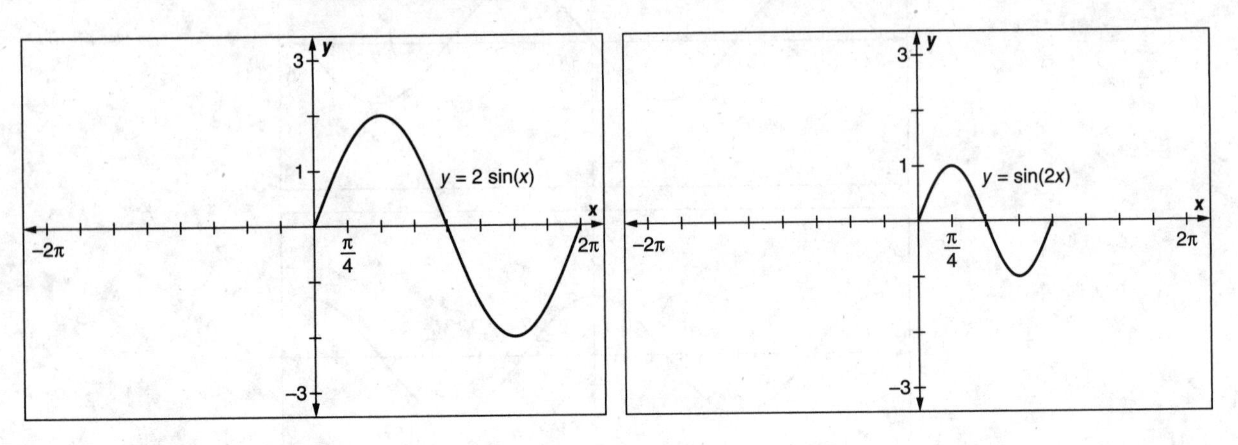

SOLUTION The graph of $y = \sin(x)$ is stretched from the x-axis by a factor of 2 to create $y = 2\sin(x)$. The graph of $y = \sin(x)$ is compressed toward the y-axis by a factor of 2 (or stretched from the y-axis by a factor of $\frac{1}{2}$) to create the graph of $y = \sin(2x)$. If you look back to the examples of $f(x) = x^2$ and $f(x) = \sqrt{x}$, you will see that each graph is stretched toward the y-axis by a factor of $\frac{1}{2}$ to create the graphs of $y = f(2x)$.

PROBLEM Describe the transformations of $y = \cos(x)$ to create the graph of $y = 3\cos(2x) + 1$. Sketch a graph of this function over one period.

SOLUTION The graph is compressed toward the y-axis by a factor of 2, stretched from the x-axis by a factor of 3, and translated up 1.

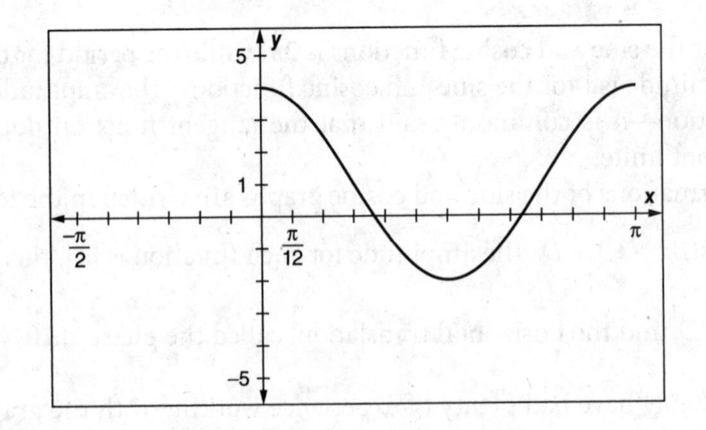

PROBLEM Describe the transformations of $y = \cos(x)$ to create the graph of $y = -3\cos\left(x + \frac{\pi}{3}\right) + 2$. Sketch a graph of this function over one period.

SOLUTION The graph is translated $\frac{\pi}{3}$ to the left (i.e., the phase shift is $\frac{\pi}{3}$ to the left), reflected over the x-axis, stretched from the x-axis by a factor of 3, and translated up 2 units.

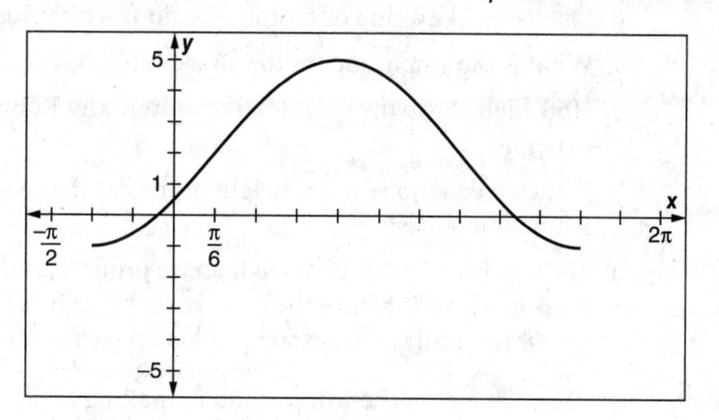

PROBLEM Describe the transformations of $y = \sin(x)$ to create the graph of $y = 2\sin\left(2x + \frac{\pi}{3}\right) + 1$. Sketch a graph of this function over one period.

SOLUTION This is the trickiest of all the problems because there is a phase shift as well as a compression toward the y-axis. The graph is compressed toward the y-axis by a factor of 2, and the phase shift is left $\frac{\frac{\pi}{3}}{2} = \frac{\pi}{6}$. The graph is stretched from the x-axis by a factor of 2, and the graph is translated up 1.

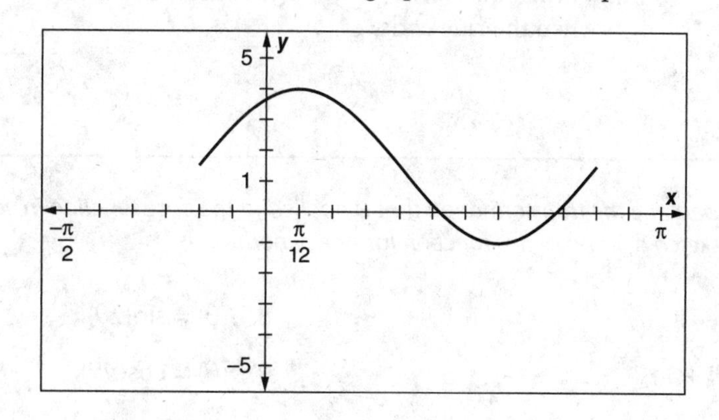

PROBLEM Andrew and Robin are at the amusement park and decide to ride the Ferris wheel. While waiting in line to get into a car on the ride, Andrew reads the following information on a board outside the control house for the Ferris wheel. "This Ferris wheel has a diameter of 100 ft and it takes 90 sec to make a complete revolution around the wheel. The center of the wheel is 60 ft above ground." Andrew and Robin are the last passengers to board before the ride begins. The height of the car that Andrew and Robin are in can be graphed as a **sinusoid** (i.e., behaves like a sine or cosine function) with respect to the time.

a. What is the amplitude for the sinusoid?

b. How high above the ground are Andrew and Robin when the ride is about to begin?

c. Write an equation for the height of the car that Andrew and Robin are in as a function of time.

d. The ride lasts 5 min. How high above ground are Andrew and Robin when the ride ends (this is before the crew goes through the motion of slowly moving the wheel from stop to stop to unload passengers)?

SOLUTION a. The diameter of the wheel is 100 ft, making the height of the wheel 50 ft. The amplitude of the sinusoid for this model is 50.

b. The bottom of the wheel is 10 ft above the ground.

c. The amplitude is 50 ft ($A = 50$), the period is 90 sec $\left(B = \frac{2\pi}{90} = \frac{\pi}{45}\right)$, and the center of the wheel is 60 ft above the ground ($D = 60$). The equation for the height, h, of their car in terms of the number of minutes, t, into the ride is $h = -50\cos\left(\frac{\pi}{45}t\right) + 60$. (Andrew and Robin are at the minimum height for this wheel/function at the beginning of the ride, which accounts for the negative coefficient.)

d. 5 min = 300 sec. $h(300) = 85$. Andrew and Robin are 85 ft above the ground when the ride ends.

EXERCISE 13·2

Describe the transformation that is applied to the base function in each question and then sketch a graph of this function for one period.

1. $f(\theta) = \sin(\theta) - 1$
2. $g(\theta) = \cos(\theta) + 1$
3. $k(\theta) = \sin\left(\theta - \frac{\pi}{2}\right)$
4. $m(\theta) = \cos\left(\theta + \frac{\pi}{3}\right)$
5. $p(\theta) = \tan\left(\theta - \frac{\pi}{4}\right)$
6. $q(\theta) = \tan(\theta) - 1$
7. $f(\theta) = 2\sin(\theta)$
8. $f(\theta) = \sin(2\theta)$
9. $f(\theta) = \cos(\pi\theta)$
10. $f(\theta) = 2\cos(2\theta) - 1$
11. $k(\theta) = \frac{-1}{2}\sin\left(\theta - \frac{\pi}{2}\right)$
12. $f(\theta) = \tan(2\theta)$
13. $g(\theta) = 2\sin\left(\theta + \frac{\pi}{4}\right) + 1$
14. $g(\theta) = -2\cos\left(\theta + \frac{\pi}{3}\right) + 3$

Determine the amplitude, period, phase shift, and vertical translation for the functions given in questions 15 and 16.

15. $g(\theta)=2\sin\left(\theta+\frac{\pi}{4}\right)+1$

16. $p(\theta)=-2\cos\left(\theta-\frac{\pi}{6}\right)+3$

Find the value of the trigonometric functions given in questions 17 and 18 and explain how you determined your answers.

17. Given a number z with $0\le z\le\frac{\pi}{2}$ and $\sin(z)=h$. What is the value of $\sin(z+2\pi)$?

18. Given a number z with $0\le z\le\frac{\pi}{2}$ and $\cos(z)=k$. What is the value of $\cos(z+3\pi)$?

Use the following information to determine answers for questions 19–21.

The Bay of Fundy, off the Atlantic coast of North America between Canada's New Brunswick and Nova Scotia, is a famous tourist site because of the difference in water depths between the low and high tides. The time difference between high and low tides is 6 hr 13 min (meaning that high tides occur every 12 hr 26 min). The depth of the water in the Bay of Fundy during high tide averages 43.95 ft. The depth of the water during low tides averages 6.85 ft. Assume that the depth of the water in the Bay of Fundy is a sinusoid.

19. Compute the amplitude for this periodic function.

20. What is the average value for this periodic function?

21. Using the period between high tides to be 12.5 hr (for the ease of computation), write an equation for the depth of the water in the Bay of Fundy. Let $t=0$ correspond to when the water is at high tide.

Inverse trigonometric functions

The graphs of $y=\sin(x)$, $y=\cos(x)$, and $y=\tan(x)$ over the interval $0\le x\le 2\pi$ are shown in the accompanying diagrams.

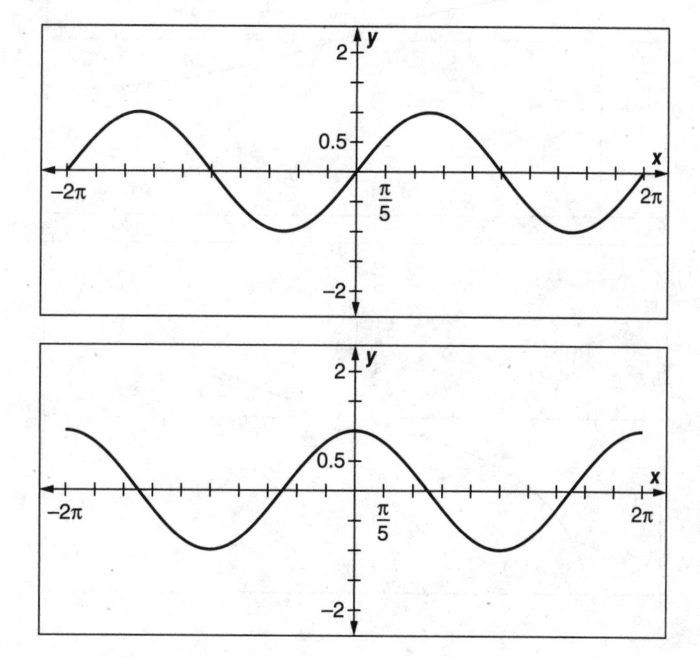

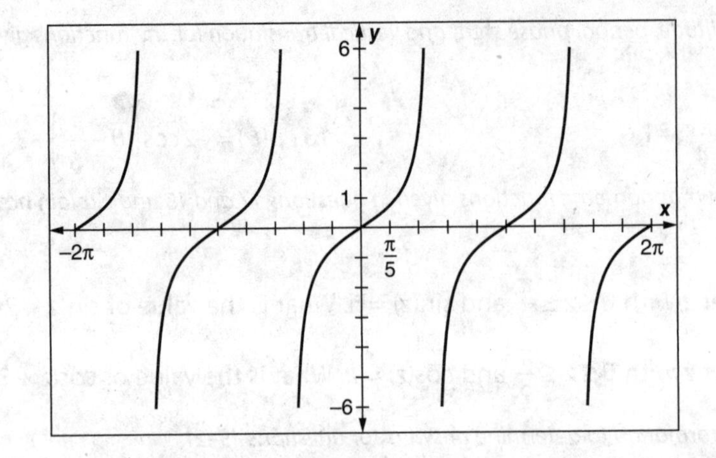

While each passes the vertical line test (indicating a function), each fails the horizontal line test (the inverse is not a function). As you have seen done with the quadratic $y = x^2$, restricting the domain to allow for a one-to-one function and covering the entire range of the function will allow for the creation of the inverse function. Convention has the restricted domain containing the value 0. The graphs of $f(x) = \sin(x)$, $g(x) = \cos(x)$, and $h(x) = \tan(x)$ over an appropriate domain are shown.

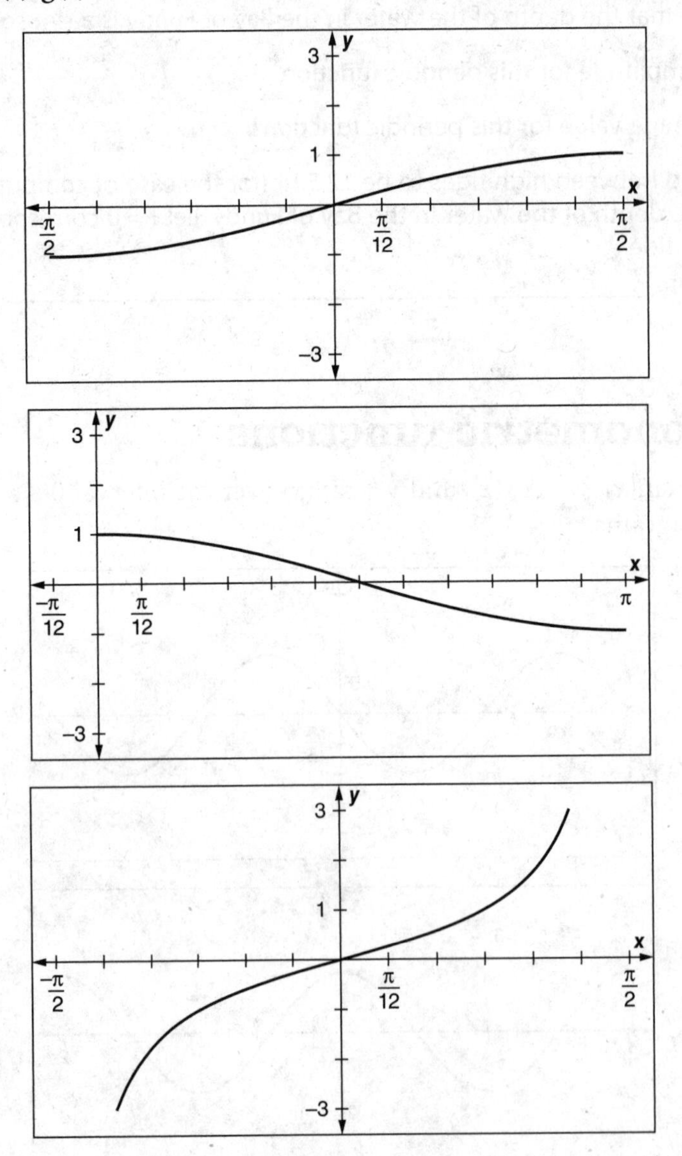

The restricted domain for sin(x) is $\frac{-\pi}{2} \le x \le \frac{\pi}{2}$, for cos($x$) is $0 \le x \le \pi$, and for tan(x) is $\frac{-\pi}{2} < x < \frac{\pi}{2}$.

It is critical that you learn these domains, because they impact the definition of the inverse trigonometric function, and provide rules for using your calculator for challenging problems.

FUNCTION	DOMAIN	RANGE (DEGREES)	RANGE (RADIANS)
$f^{-1}(x) = \sin^{-1}(x)$[or arcsin(x)]	$-1 \le x \le 1$	$-90 \le y \le 90$	$\frac{-\pi}{2} \le y \le \frac{\pi}{2}$
$g^{-1}(x) = \cos^{-1}(x)$[or arccos(x)]	$-1 \le x \le 1$	$0 \le y \le 180$	$0 \le y \le \pi$
$h^{-1}(x) = \tan^{-1}(x)$[or arctan(x)]	Reals	$-90 < y < 90$	$\frac{-\pi}{2} < y < \frac{\pi}{2}$

While it is true that $\sin(240) = \sin(300) = \frac{-\sqrt{3}}{2}$, $\sin^{-1}\left(\frac{-\sqrt{3}}{2}\right) = -60$ is the only acceptable answer, because $\sin^{-1}(x)$ is a function and, as such, can have only one output value for a given input.

PROBLEM Find the exact value of $\cos(\sin^{-1}(5/13))$.

SOLUTION $\sin^{-1}\left(\frac{5}{13}\right)$ is an angle. If you call the angle A, the equation $\sin(A) = \frac{5}{13}$ is equivalent, as long as A is between 0 and 90°. The problem is now the equivalent of looking for the cosine of this angle. To complete the problem, find the adjacent side of a right triangle in which the opposite side is 5 and the hypotenuse is 13. Use the Pythagorean theorem to determine that the adjacent side is 12. Therefore, $\cos(A) = \frac{12}{13}$.

This problem can also be solved using the Pythagorean identity $\cos^2(A) + \sin^2(A) = 1$, and using only the positive solution to the equation because of the restriction on angle A.

PROBLEM Find the exact value of $\tan\left(\sin^{-1}\left(\frac{\sqrt{7}}{4}\right)\right)$.

SOLUTION Let $\sin^{-1}\left(\frac{\sqrt{7}}{4}\right) = A$, so $\sin(A) = \frac{\sqrt{7}}{4}$. Use the Pythagorean theorem to solve $x^2 + (\sqrt{7})^2 = 4^2$ to determine that x, the adjacent side of the triangle, is 3. $\tan(A) = \frac{\sqrt{7}}{3}$.

PROBLEM Find the exact value of $\tan\left(\cos^{-1}\left(\frac{-\sqrt{3}}{5}\right)\right)$.

SOLUTION As before, let $\cos^{-1}\left(\frac{-\sqrt{3}}{5}\right) = A$, so $\cos(A) = \frac{-\sqrt{3}}{5}$. (Note that $90° < A < 180°$.) Find the length of the opposite side of the triangle using the equation $(-\sqrt{3})^2 + y^2 = 5^2$ to determine that $y = \sqrt{22}$. (The y-coordinate in the second quadrant is positive, so do not use the negative solution to this equation.) $\tan(A) = \frac{\sqrt{22}}{-\sqrt{3}} = \frac{\sqrt{66}}{-\sqrt{3}}$.

PROBLEM Find the exact value of $\sin\left(\tan^{-1}\left(\frac{x}{2}\right)\right)$.

SOLUTION Although having the variable in the problem might make the problem look different, it is not. Let $\tan^{-1}\left(\frac{x}{2}\right) = A$, so $\tan(A) = \frac{x}{2}$. The legs of the right triangle have lengths x and 2, so the hypotenuse has length $\sqrt{x^2+4}$ and $\sin(A) = \frac{x}{\sqrt{x^2+4}} = \frac{x\sqrt{x^2+4}}{x^2+4}$. Observe that this is the correct answer whether x is positive or negative.

PROBLEM Find the exact value of $\cos\left(\sin^{-1}\left(\frac{2}{3}\right) + \tan^{-1}\left(\frac{4}{5}\right)\right)$.

SOLUTION Let $A = \sin^{-1}\left(\frac{2}{3}\right)$ and $B = \tan^{-1}\left(\frac{4}{5}\right)$, so that $\sin(A) = \frac{2}{3}$ and $\tan(B) = \frac{4}{5}$. Working with $\sin(A) = \frac{2}{3}$ first, solve $x^2 + 2^2 = 3^2$ to find $x = \sqrt{5}$ and $\cos(A) = \frac{\sqrt{5}}{3}$. Working with $\tan(B) = \frac{4}{5}$, $4^2 + 5^2 = c^2$ gives $c = \sqrt{41}$ and $\sin(B) = \frac{4}{\sqrt{41}} = \frac{4\sqrt{41}}{41}$, while $\cos(B) = \frac{5}{\sqrt{41}} = \frac{5\sqrt{41}}{41}$. The original problem can be written as $\cos(A + B)$ which, by the identity, equals $\cos(A)\cos(B) - \sin(A)\sin(B)$. Substituting the values found, $\cos(A)\cos(B) - \sin(A)\sin(B) = \left(\frac{\sqrt{5}}{3}\right)\left(\frac{5\sqrt{41}}{41}\right) - \left(\frac{2}{3}\right)\left(\frac{4\sqrt{41}}{41}\right)$. Therefore,

$$\cos\left(\sin^{-1}\left(\frac{2}{3}\right) + \tan^{-1}\left(\frac{4}{5}\right)\right) = \frac{5\sqrt{205} - 8\sqrt{41}}{123}.$$

EXERCISE

13·3

Find the exact value of the trigonometric functions given in questions 1 and 2.

1. $\tan\left(\sin^{-1}\left(\frac{-\sqrt{5}}{4}\right)\right)$

2. $\sin\left(\tan^{-1}\left(\frac{-5}{7}\right)\right)$

Find an algebraic expression for the function given in question 3.

3. $\cos\left(2\tan^{-1}\left(\frac{x}{3}\right)\right)$

Find the exact value of the trigonometric functions given in questions 4 and 5.

4. $\sin\left(2\tan^{-1}\left(\frac{3}{7}\right)\right)$

5. $\cos\left(\sin^{-1}\left(\frac{2}{3}\right)+\sin^{-1}\left(\frac{-1}{4}\right)\right)$

Solving trigonometric equations

Solving trigonometric equations requires that you know the special angles, can use a calculator to determine reference angles, and can recognize identities.

PROBLEM Solve $4\cos(A)+3=1$ $(0 \leq A < 360°)$.

SOLUTION Subtract 3 and divide by 4 to get $\cos(A)=\frac{-1}{2}$. Because $\cos(60)=\frac{1}{2}$, use 60° as the reference angle for the second and third quadrants. $A = 120°, 240°$.

PROBLEM Solve $4\cos(A)+2=1$ $(0 \leq A < 360°)$, to the nearest tenth of a degree.

SOLUTION Subtract 2 and divide by 4 to get $\cos(A)=\frac{-1}{4}$. This does not represent a special angle, so the calculator will be necessary. While it is true that the calculator can compute $\cos^{-1}\left(\frac{-1}{4}\right)=104.5°$, you will still need to find the reference angle $(180° - 104.5° = 75.5°)$ to find the third quadrant angle $(180° + 75.5° = 255.5°)$. $A = 104.5°, 255.5°$.

PROBLEM Solve $\tan^2(A) - 8 = 0$ $(0 < A < 360°)$ to the nearest tenth of a degree.

SOLUTION Add 8 and take the square root of both sides to find that $\tan(A) = \pm 8$. The reference angle for this problem is $\tan^{-1}(\sqrt{8}) = 70.5°$. $\tan(A) > 0$ in quadrant III $(180° + 70.5° = 250.5°)$. $\tan(A) < 0$ in quadrant II $(180° - 70.5° = 109.5°)$, and quadrant IV $(360° - 70.5° = 289.5°)$. $A = 70.5°, 109.5°, 250.5°, 289.5°$.

PROBLEM Solve $4\sin^2(A) - 3\sin(A) - 1 = 0$ $(0 \leq A < 360°)$, to the nearest tenth of a degree.

SOLUTION The equation is a quadratic with the variable being the trigonometric function, $\sin(A)$, rather than just a variable. (If it is helpful, replace $\sin(A)$ with y and solve the equation.) The quadratic factors to $[4\sin(A) + 1][\sin(A) - 1]$.

Solve for each factor: $\sin(A) = \frac{-1}{4}, 1$

$\sin(A) = 1$ when $A = 90°$.

The reference angle for $\sin(A) = \frac{-1}{4}$ is $\sin^{-1}\left(\frac{1}{4}\right) = 14.5°$. $\sin(A) < 0$ in quadrant III ($180° + 14.5° = 194.5°$) and in quadrant IV ($360° - 14.5° = 345.5°$). $A = 90°, 194.5°, 345.5°$.

PROBLEM Solve $\sin(2A) = \cos(A)$ ($0 \le A < 360°$).

SOLUTION $\sin(2A) = 2\sin(A)\cos(A)$, so the equation becomes $2\sin(A)\cos(A) = \cos(A)$. The temptation is to divide by $\cos(A)$—don't do it. Because $\cos(A)$ could equal zero, you shouldn't divide. Rather (as you have always done with variable expressions) subtract $\cos(A)$ to get $2\sin(A)\cos(A) - \cos(A) = 0$.

Factor: $\cos(A)\,[2\sin(A) - 1] = 0$

Solve: $\cos(A) = 0$ or $\sin(A) = \frac{1}{2}$

$\cos(A) = 0$ when $A = 90°$ or $270°$, and $\sin(A) = \frac{1}{2}$ when $A = 30°$ or $150°$. Therefore, $A = 30°, 90°, 150°, 270°$.

PROBLEM Solve $\cos(2A) + \sin(A) = 1$ ($0 \le A < 360°$).

SOLUTION There are three different identities for $\cos(2A)$. Because the other function in the problem is $\sin(A)$, use the identity $1 - 2\sin^2(A)$ for $\cos(2A)$. The equation now becomes $1 - 2\sin^2(A) + \sin(A) = 1$, or $-2\sin^2(A) + \sin(A) = 0$. Factor $\sin(A)$ $[-2\sin(A) + 1] = 0$, so $\sin(A) = 0$ or $\sin(A) = \frac{1}{2}$. $\sin(A) = 0$ when $A = 0°$ or $180°$, and $\sin(A) = \frac{1}{2}$ when $A = 30°$ or $150°$. Therefore, $A = 0°, 30°, 150°, 180°$.

PROBLEM Solve $6\tan^2(A) - 5\tan(A) - 3 = 0 (0 \le A < 360°)$.

SOLUTION Because this quadratic will not factor, the quadratic formula will be needed. Caution: The variable of this quadratic is $\tan(A)$, not A.

$$\tan(A) = \frac{-(-5) \pm \sqrt{(-5)^2 - 4(6)(-3)}}{2(6)} = \frac{5 \pm \sqrt{25 + 72}}{12} = \frac{5 \pm \sqrt{97}}{12}.$$

$$\tan(A) = \frac{5 + \sqrt{97}}{12} \text{ or } \tan(A) = \frac{5 - \sqrt{97}}{12}.$$

With equations like these that have the radical expression, it is worthwhile to type $\left(\frac{5 + \sqrt{97}}{12}\right)$ into your calculator and store the decimal in one of the calculator's memory locations (e.g., the TI-84 uses the STO button and any of the letters for memory locations; the TI-Nspire allows for the storage of a value on the calculator screen with the = command). Use the inverse tangent function with the memory location [e.g., $\tan^{-1}(B)$]. Do not enter the decimal yourself—you can mistype a long decimal or inappropriately round the decimal to too few digits.

When $\tan(A) = \frac{5+\sqrt{97}}{12}$, the reference angle is $\tan^{-1}\left(\frac{5+\sqrt{97}}{12}\right) = 51.1°$. $\tan(A) > 0$ in quadrant III as well ($180° + 51.1° = 231.1°$). When $\tan(A) = \left(\frac{5-\sqrt{97}}{12}\right)$, the reference angle is $\tan^{-1}\left(-\left(\frac{5-\sqrt{97}}{12}\right)\right) = \tan^{-1}\left(\frac{\sqrt{97}-5}{12}\right) = 22.0°$. $\tan(A) < 0$ in quadrant II ($180° - 22.0° = 158.0°$) and quadrant IV ($360° - 22.0° = 338.0°$). Therefore, $A = 51.1°, 158.0°, 231.1°, 338.0°$.

PROBLEM As you saw in an earlier section of this chapter, Andrew and Robin determined that their height above ground as they rode the Ferris wheel is given by the formula $h = -50\cos\left(\frac{\pi}{45}t\right) + 60$. At what times during the first 90 sec were they 90 ft off the ground?

SOLUTION With $h = 90$, the problem becomes $90 = -50\cos\left(\frac{\pi}{45}t\right) + 60$. Subtract 60 and divide by -50 to get $\cos\left(\frac{\pi}{45}t\right) = \frac{-3}{5}$. A major part of this problem is to recognize that the problem uses radian mode—this is usually the case when working with real-life applications. The reference angle is $\cos^{-1}\left(\frac{3}{5}\right) = 0.927$. $\cos(A) < 0$ in quadrant II ($\pi - 0.927 = 2.2143$) and quadrant III ($\pi + 0.927 = 4.0689$). Therefore, $\frac{\pi}{45}t = 2.2143$ and $\frac{\pi}{45}t = 4.0689$. Solve each of these values to get $t = 31.7$ and 58.3. Andrew and Robin will be 90 ft above ground 31.7 sec after the ride starts going up, and again 58.3 sec after the ride starts coming down.

EXERCISE
13·4

Solve each of the following equations for $0 \le A < 360°$. Round answers to the nearest tenth of a degree when necessary.

1. $4\sin(A) + 3 = 0$
2. $5\tan(A) - 9 = 0$
3. $2\cos^2(A) - 3\cos(A) + 1 = 0$
4. $3\sin^2(A) + 5\sin(A) - 2 = 0$
5. $7\tan^2(A) + 8\tan(A) - 3 = 0$
6. $\sin(2A) = \sin(A)$
7. $\cos(2A) + \cos(A) = 1$
8. $2\cos(2A) - \sin(A) - 1 = 0$
9. $2\cos(A) - \sec(A) = 1$
10. $\sin(2A) = \tan(A)$

Use the following information to determine the answer to question 11.

The depth of the water in the Bay of Fundy is given by the equation $d = 18.55\cos\left(\frac{4\pi}{25}t\right) + 25.4$, with $t = 0$ corresponding to high tide.

11. How many hours after high tide is the depth of the water 30 ft?

Answer key

1 Functions: An introduction

1·1

1. $D_A = \{-4, -2, 3, 5\}$
2. $R_A = \{0, 1, 4, 9\}$
3. $D_B = \{-9, -7, 2, 3, 5, 6\}$
4. $R_B = \{-2, -1, 0, 1, 3\}$
5. $D_C = \{-5, -2, 3, 4, 9\}$
6. $R_C = \{-1, 0, 1, 2, 3\}$
7. $A^{-1} = \{(4, -2), (1, 5), (0, -4), (9, -4), (1, 3)\}$
8. $B^{-1} = \{(3, -7), (0, 2), (1, -9), (3, 3), (-2, 6), (-1, 5)\}$
9. $C^{-1} = \{(1, -2), (2, 3), (-1, 4), (3, -5), (0, 9)\}$

1·2

1. B and C
2. C^{-1}
3. No. Many students have cell phones as well as land lines at home at which they can be reached.
4. No, calling a home land line may reach the student or any other member of the family.
5. Yes, a person may legally have only one Social Security number.
6. Yes, each Social Security number issued is unique to a given person.
7. Yes, a person only has one birthday.
8. Possibly not. Does anyone in your class have the same birthday?

1·3

1. 20
2. -7
3. $-3n + 2$
4. 11
5. 1/2
6. $\dfrac{2t-1}{t-5}$
7. 5
8. 1
9. $\sqrt{4r-3}$

1·4

1. 6
2. $\sqrt{7}$
3. -1
4. 4
5. 104
6. -4
7. $\{x|x \neq 4\}$
8. $\{x|x \neq -1/5, 1\}$
9. $x \geq -2$
10. $x \leq 10$

1·5

1. The graph of the parabola moves left 2 units, is stretched from the x-axis by a factor of 3 units, and slides down 1.
2. The graph of the absolute value slides to the right 1 unit and down 3 units.

3. The graph of the square root function slides to the left 1 unit and up 2 units.
4. The graph of the parabola is reflected over the x-axis, stretched from the x-axis by a factor of 2, and slides up 3 units.
5. The graph of the absolute value function slides to the left 2 units, is reflected over the x-axis, stretched from the x-axis by a factor of 1/3, and slides up 5 units.

1·6

1. $f^{-1}(x)=\dfrac{x+5}{3}$
2. $g^{-1}(x)=\dfrac{x-5}{-8}$ or $\dfrac{5-x}{8}$
3. $k^{-1}(x)=\dfrac{2x+7}{x-3}$
4. $p^{-1}(x)=\dfrac{-4x-1}{3x-5}$

1·7

1. b, d
2. a, d

2 Linear equations and inequalities

2·1

1. $a=-3$
2. $y=3.7$
3. $x=6$
4. $g=\dfrac{-5}{8}$
5. $a=\dfrac{t}{7}$
6. $q=24.5r$
7. Kristen has 445 classical and 3225 rock songs on her MP3 player.
8. Diane bought 25 plants at \$6.50 each and 65 plants at \$4.50 each.

2·2

1. $(4, -3)$
2. $(-5, 1)$
3. $(4.5, 6.5)$
4. $(-2, -5)$
5. $(7, 0)$
6. $(-1, 8)$
7. $\left(\dfrac{-71}{7}, \dfrac{69}{7}\right)$
8. $(12, -5)$

2·3

1. $(3, -7)$
2. $(-5, 2)$
3. $(-10, 12)$
4. $(6.3, 5.2)$
5. $(-12.2, 13.8)$

2·4

1. $(7, -3)$
2. $(-8, 5)$
3. $(120, 135)$
4. $\left(\dfrac{-5}{6}, \dfrac{2}{3}\right)$
5. $(8.4, 4.2)$
6. $\left(2\dfrac{1}{2}, 4\dfrac{1}{6}\right)$

2·5

1. $(2, -3, 4)$
2. $(5, 2, -4)$
3. $(7, 0, -2)$
4. $\left(\dfrac{5}{2}, \dfrac{-2}{3}, \dfrac{-3}{4}\right)$

2·6

1. (5, −8)
2. (−3, 11)
3. $\left(\frac{-1}{6}, \frac{3}{4}\right)$
4. (40, 26, 85)
5. (3, 0, 2)
6. $\left(\frac{2}{3}, \frac{3}{5}, \frac{-1}{4}\right)$
7. (2, 5, 9, 4)
8. (10, −5, 8, −4)

2·7

1. Advance sale $7.50; at the door $9; adult $15
2. 250 pre-performance student tickets; 300 student tickets on the day of the performance; 700 adult tickets
3. B is worth 17 points, A is worth 13 points, E is worth 11 points, and R is worth 19 points.

2·8

1. $x < 6$

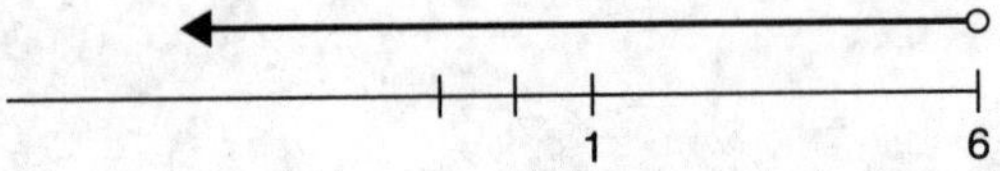

2. $x \geq 15$

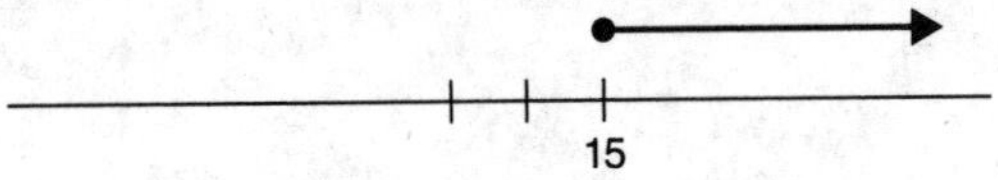

3. $9 \leq x < 13$

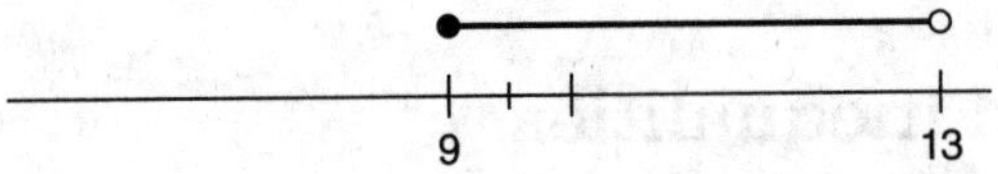

4. $-1 \leq x < 4$

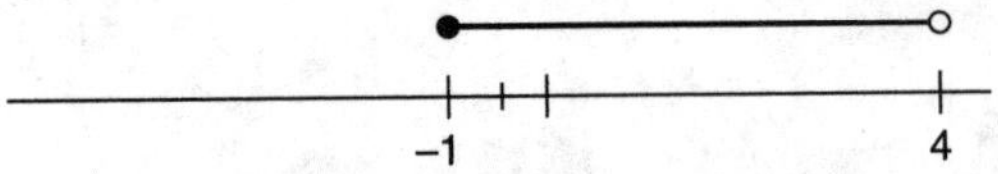

5. $x < \frac{14}{3}$ or $x \geq 5$

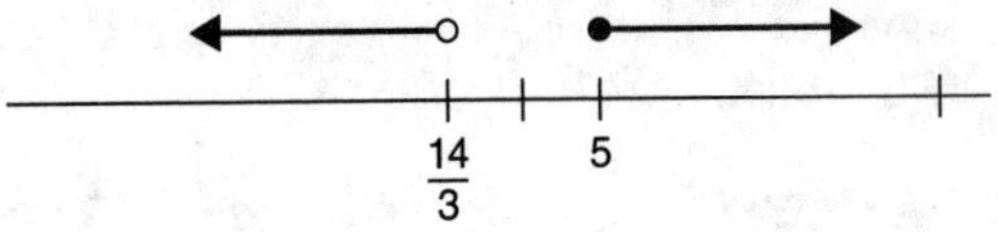

6. All real numbers

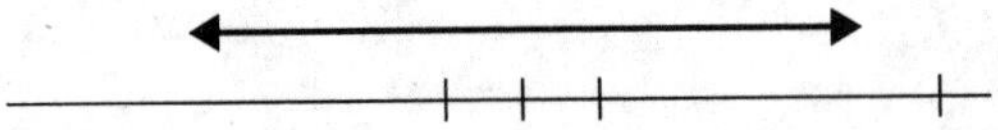

2·9

1.

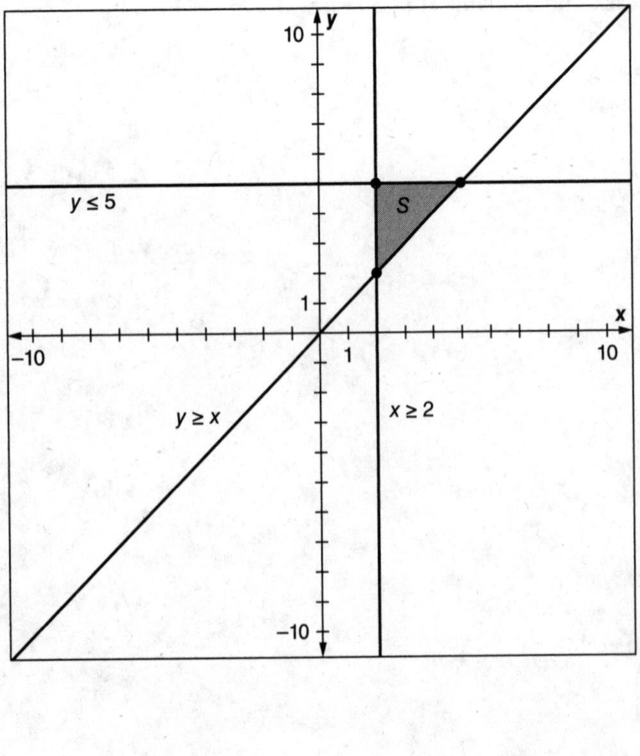

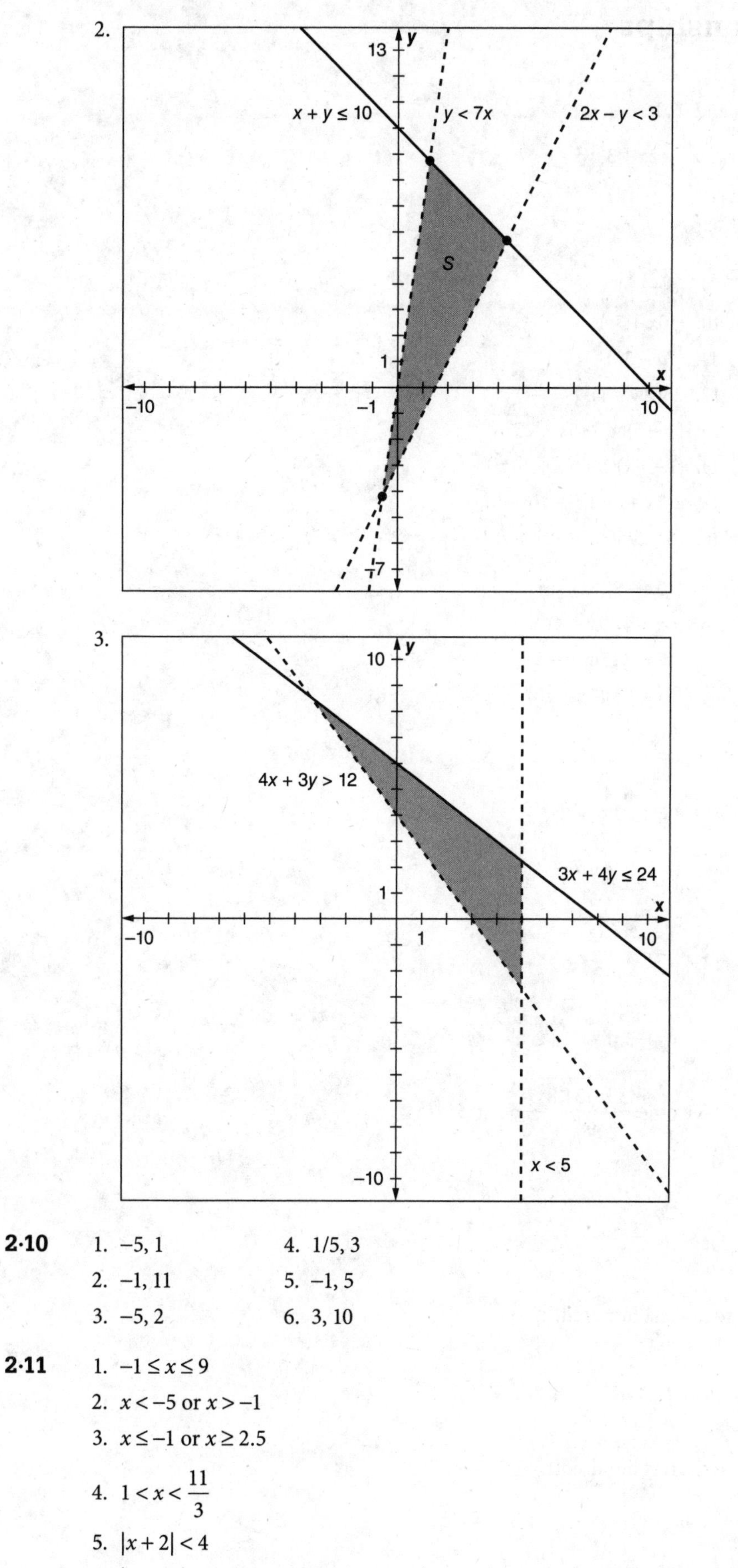

2·10

1. $-5, 1$
2. $-1, 11$
3. $-5, 2$
4. $1/5, 3$
5. $-1, 5$
6. $3, 10$

2·11

1. $-1 \le x \le 9$
2. $x < -5$ or $x > -1$
3. $x \le -1$ or $x \ge 2.5$
4. $1 < x < \dfrac{11}{3}$
5. $|x + 2| < 4$
6. $|2x - 5| \ge 7$ (This is equivalent to $|x - 2.5| \ge 3.5$ but uses integers in its expression.)

3 Quadratic relationships

3·1

1. $x = -2$
2. $(-2, -17)$
3. $y \geq -17$
4. $x = 4.5$
5. $(4.5, 17.5)$
6. $y \leq 17.5$
7. $(5, 0)$ and $(6,0)$
8. $p(x) = \frac{1}{4}x^2 - 2x - 3$
9. $(-3, -1)$
10. $x = -3$

3·2

1. $8x(x - 2)$
2. $15(x^2 + 16)$
3. $4x^2 (2x^2 - 3x + 5)$
4. $(4x - 3)(8x^2 + 3x - 9)$
5. $(x + y)(x - y)$
6. $(7y - 5)(7y + 5)$
7. $9(3t + 4)(3t - 4)$
8. $(3x + 2y + 3)(3x - 2y + 5)$
9. $(a + 9b)^2$
10. $(10x + 7y)^2$
11. $(7a - 3b)^2$
12. $25(2w - v)^2$
13. $(x + 5)(x^2 - 5x + 25)$
14. $(1 - 4x)(1 + 4x + 16x^2)$
15. $2(2x + 3)(4x^2 - 6x + 9)$
16. $(6 - 5k)(36 + 30k + 25k^2)$

3·3

1. $(3x - 5)(4x + 11)$
2. $(8x + 9)(6x + 7)$
3. $(6x + 13)(4x - 5)$
4. $(12x + 7)(3x + 4)$
5. $(10x - 9)(5x - 6)$
6. $(6x + 5)(9x + 4)$
7. $(12x - 25)(4x + 9)$
8. $(8x - 15)(3x + 10)$

3·4

1. $f(x) = (x - 3)^2 - 4$
2. $g(x) = 5\left(x + \frac{3}{2}\right)^2 - \frac{57}{4}$
3. $p(x) = \frac{1}{4}(x + 10)^2 - 31$
4. $q(x) = \frac{-2}{5}(x - 5)^2 + 13$
5. $x = -4 \pm \sqrt{21}$
6. $x = \frac{-5}{2}, \frac{1}{2}$
7. $x = 4 \pm \frac{4\sqrt{6}}{3}$
8. $x = \frac{-3}{2}, \frac{-2}{3}$

3·5

1. $x = \frac{9 \pm \sqrt{129}}{8}$
2. $x = \frac{8 \pm \sqrt{78}}{2}$
3. $x = \frac{12 \pm 9\sqrt{2}}{2}$
4. $x = -2, 14$
5. $x = \frac{143 \pm \sqrt{31{,}817}}{98}$

3·6

The function is $s(t) = -16t^2 + 64t + 40$

1. 104 ft
2. 0.65 and 3.35 sec (answer to nearest hundredth)
3. 4.55 sec

The function is $s(t) = -4.9t^2 + 19.6t + 15$

4. 34.6 m
5. 1.03 and 2.97 sec (answer to nearest hundredth)
6. 4.66 sec
7. 600 items
8. $360,000

9. \$75,000
10. \$285,000
11. $P(x) = -x^2 + 1100x - 15000$
12. The company begins to make money when between 13.81 and 1086.19 units are sold. It loses money when less than 13.81 or more than 1086.19 units are sold.
13. 550
14. 287,500

3·7

1. Domain: $x \geq -3$; inverse function: $y = \sqrt{2x+4} - 3$
2. Domain: $x \geq 5$; inverse function: $y = \sqrt{\frac{-1}{3}x + \frac{2}{3}} + 5$

3·8

1. $(x+5)^2 + (y-2)^2 = 82$
2. Center: $(-4, 9)$; $r = 9$
3. Center: $(6, 0)$; $r = 3\sqrt{2}$
4. Center: $(-7, 10)$; $r = 13$
5. Center: $(21, -60)$; $r = 71$

3·9

1. $\frac{(x+2)^2}{9} + \frac{(y+1)^2}{36} = 1$
2. $\left(\frac{-2 \pm 3\sqrt{2}}{2}, 0\right), \left(0, \frac{-2}{3}\right), \left(0, \frac{14}{3}\right)$
3. $\frac{(x-7)^2}{25} + \frac{(y+3)^2}{9} = 1$
4. $\frac{(x+3)^2}{50} + \frac{(y-4)^2}{30} = 1$

3·10

1. Center: $(4, -5)$; vertices: $(4 \pm \sqrt{6}, -5)$; asymptotes: $y + 5 = \pm 2(x - 4)$
2. Center: $(-5, 4)$; vertices: $(-5, 4 \pm \sqrt{6})$; asymptotes: $y - 4 = \pm \frac{1}{2}(x + 5)$
3. $\frac{(y-7)^2}{16} - \frac{(x-1)^2}{49} = 1$
4. $\frac{(x+2)^2}{64} - \frac{(y-9)^2}{81} = 1$

3·11

1. $(\sqrt{2}, \sqrt{2} - 5), (-\sqrt{2}, \sqrt{2} - 5)$
2. $(-7, -1), (1, 7)$
3. $\left(\frac{2\sqrt{55}}{5}, \frac{9\sqrt{5}}{5}\right), \left(\frac{2\sqrt{55}}{5}, \frac{-9\sqrt{5}}{5}\right), \left(\frac{-2\sqrt{55}}{5}, \frac{-9\sqrt{5}}{5}\right), \left(\frac{-2\sqrt{55}}{5}, \frac{9\sqrt{5}}{5}\right)$

4 Complex numbers

4·1

1. i
2. -1
3. i
4. i
5. -64
6. $5i\sqrt{2}$
7. -45
8. $-4i\sqrt{30}$

4·2

1. $13+2i$
2. $10-2i$
3. $23-2i$
4. $58-2i\sqrt{5}$
5. $4+42i\sqrt{5}$
6. $\frac{9}{5}-\frac{12}{5}i$
7. $\frac{7}{4}+\frac{\sqrt{3}}{12}i$
8. $x=3, y=-2$

4·3

1. Discriminant = 0; roots are real, rational, and equal
2. Discriminant = 204; roots are real, irrational, and unequal
3. Discriminant = −79; roots are complex numbers
4. Discriminant = 4; roots are real, rational, and unequal
5. $x=\frac{1}{2}\pm\frac{3}{4}i$
6. $x=\frac{3}{8}\pm\frac{\sqrt{5}}{4}i$

4·4

1. Sum $=\frac{-7}{2}$; product $=\frac{-3}{2}$
2. 3
3. 5
4. $36x^2-17x-35=0$
5. $81x^2-270x+127=0$
6. $64x^2-80x+97=0$

5 Polynomial functions

5·1

1. Even
2. Neither
3. Odd
4. Neither
5. Even

5·2

1. $x\leq\frac{1}{2}$
2. Reals
3. 5
4. −7

5·3

1. $x\to-\infty, f(x)\to\infty; x\to\infty, f(x)\to-\infty$
2. $x\to-\infty, g(x)\to-\infty; x\to\infty, g(x)\to-\infty$
3. $x\to-\infty, k(x)\to-\infty; x\to\infty, k(x)\to\infty$
4. $x\to-\infty; p(x)\to\infty; x\to\infty, p(x)\to\infty$

5·4

1. Yes
2. Yes
3. $(4x+3)(5x-4)(3x+7)$
4. $(5x+3)(3x-4)(2x+7)(8x-15)$

6 Rational and irrational functions

6·1

1. $x \neq 2, 3$
2. $x = \frac{-7}{5}$
3. $y \to 0$
4. $x \neq -7, \frac{2}{3}$
5. $x = -4, \frac{3}{2}$
6. $y \to \frac{2}{3}$
7. $x \neq \frac{3}{4}$
8. $x = \frac{-8}{9}$
9. $y \to \frac{9}{4}$
10. $k^{-1}(x) = \frac{3x+8}{4x-9}$

6·2

1. $\frac{3x+2}{4x+1}$
2. $\frac{8x-3}{2x-3}$
3. $\frac{-4(x-1)}{5}$
4. $\frac{x^2+x+1}{2x-3}$
5. $\frac{3x-5}{2x+1}$
6. $\frac{-3(x+4)}{x-4}$

6·3

1. $\frac{x-5}{3x+2}$
2. $\frac{x+5}{5x}$
3. $\frac{3x+2}{2x+3}$
4. $\frac{x}{2}$
5. $\frac{x^2+9x+16}{2(x+5)(x-5)}$
6. $\frac{5x^2-13x-1}{(x-3)(x-2)(2x+3)}$
7. $\frac{10x^2+21x-37}{6(5x-1)(5x+1)}$
8. $\frac{3x^2-16x-1}{(2x+1)(3x-1)(3x+1)}$
9. $\frac{2x}{x^2-3}$
10. $\frac{x}{x-1}$

6·4

1. $\frac{3}{2}, 6$
2. $-1, 27$
3. $4, 12$
4. $4, \frac{1}{2}$
5. $4, \frac{21}{2}$
6. -4
7. 40, 60, 120 farads

6·5

1. $x \geq \frac{4}{3}$
2. $y \geq 5$
3. $x \leq \frac{2}{3}$
4. $y \leq 7$
5. $x \leq 3$ or $x \geq 9$
6. $y \geq 0$
7. $-8 \leq x \leq 4$
8. $0 \leq y \leq 6$

6·6

1. $5\sqrt{7}$
2. $10\sqrt{2}$
3. $10\sqrt[3]{2}$
4. 6
5. $\frac{2\sqrt{3}}{3}$
6. $2(4-\sqrt{3})$
7. $\frac{13-2\sqrt{2}}{7}$

6·7

1. 26
2. −15
3. 9
4. −9, −8
5. 12
6. 15

7 Exponential and logarithmic functions

7·1

1. $10x^5y^5z^5$
2. $250x^7$
3. $2000x^7$
4. $\frac{2}{3x^2}$
5. $\frac{2}{3x^2y^2}$
6. 1
7. $2x^3$
8. 27
9. $\frac{16}{9}$
10. $\frac{2x^4}{3y}$
11. $\frac{9}{100x^4y^4}$
12. $10y+15$

7·2

1. $\log_9(6561)=4$
2. $\log_{\frac{2}{3}}\left(\frac{81}{16}\right)=-4$
3. $10^2=100$
4. $3^{-3}=\frac{1}{27}$
5. Domain: $x<\frac{4}{5}$; Range: real numbers
6. Translate right 4; stretch from the x-axis by a factor of 2; translate up 1

7·3

1. $\log_b\left(\frac{(x+3)^4\sqrt{x+4}}{2x+1}\right)$
2. $\log_b\left(\frac{x+7}{(x-1)^2(x^2+1)}\right)$
3. $\log_b\left(\frac{(x+7)(x+1)}{(2x-1)\sqrt[3]{(x+9)^2}}\right)$
4. $\frac{1}{2}x$
5. $\frac{1}{2}y$
6. $\frac{3}{2}y$
7. $x+y+z$
8. $y-x$
9. $\frac{1}{2}x+\frac{1}{4}y+z$
10. $x+\frac{1}{2}y-z$

7·4

1. $\frac{1}{5}$
2. $\frac{4}{7}$
3. ± 13
4. 5
5. 1.83
6. 0.76
7. a. 0.003981
   b. 0.00000001
8. Lemon juice is an acid and sea water is a base.

8 Sequences and series

8·1

1. $7 + 11 + 15 + 19 + 23$
2. $2 + 9 + 28 + 65$
3. $85 + 90 + 95 + 100$
4. $4 + 10 + 28 + 82$
5. $\sum_{n=1}^{6}(6n-1)$
6. $\sum_{n=0}^{4}(120-5n)$ or $\sum_{n=1}^{5}(125-5n)$
7. $\sum_{n=0}^{9}(8n+81)$ or $\sum_{n=1}^{10}(8n+73)$
8. $\sum_{n=1}^{11}(2^n)$

8·2

1. 16, 25, 34, 43, 52
2. 72, 66, 60, 54, 48
3. 20, 100, 500, 2500, 12,500
4. 7, 22, 67, 202, 607
5. 17, 28, 45, 73, 118
6. $\frac{1}{2}$, 32, $\frac{1}{128}$, 131,072, $\frac{1}{2147483648}$
7. 20

8·3

1. 641
2. 1588
3. 1196
4. 209
5. 719
6. 814

8·4

1. 15,480
2. 4220
3. 11,880
4. 102,375
5. 6480
6. 25,710

8·5

1. 4,718,592
2. $\frac{2048}{2187}$
3. $\frac{4,782,969}{2048}$
4. $\frac{625}{512}$
5. 8,388,608
6. \$14,482.98

8·6

1. 4,194,300
2. 34,867,844,000
3. 240,000
4. $\dfrac{8}{11}$
5. 98,301

9 Introduction to probability

9·1

1. 30
2. 10,368 (repetition of flavors is allowed)
3. 744
4. 362,916
5. 5040
6. $n^2 - n$
7. (7!)(.25 min) = 1260 min = 21 hr (The guys probably didn't take all the pictures.)

9·2

1. 15,120
2. 151,200
3. 5040
4. 1260
5. 42,840
6. 95,040
7. $\dfrac{(1)({}_{11}P_4)}{95{,}040} = \dfrac{1}{12}$
8. The order could be Will, Peter, and seven other batters; or Peter, Will, and seven other batters. There are two ways in which Will and Peter can be arranged and ${}_{13}P_7$ ways the remaining players can be placed in the line-up. The probability is $\dfrac{(2)({}_{13}P_7)}{{}_{15}P_9} = \dfrac{1}{105}$.

9·3

1. 210
2. 495
3. ${}_{22}C_7 = 170{,}544$
4. $\dfrac{({}_{12}C_4)({}_{10}C_3)+({}_{12}C_5)({}_{10}C_2)+({}_{12}C_6)({}_{10}C_1)+({}_{12}C_7)({}_{10}C_0)}{{}_{22}C_7} = \dfrac{199}{323}$
5. $\dfrac{({}_{2}C_2)({}_{20}C_5)}{{}_{22}C_7} = \dfrac{1}{11}$
6. Five of the cards contain lengths that represent right triangles (3, 4, 5; 5, 12, 13; 1, $\sqrt{3}$, 2; 11, 60, 61; 7, 24, 25). Selecting three cards from these five and one card from the remaining three has probability $\dfrac{({}_{5}C_3)({}_{3}C_1)}{{}_{8}C_4} = \dfrac{3}{7}$.

9·4

1. $32x^5 + 80x^4y + 80x^3y^2 + 40x^2y^3 + 10xy^4 + y^5$
2. $x^6 - 12x^5y + 60x^4y^2 - 160x^3y^3 + 240x^2y^4 - 192xy^5 + 64y^6$
3. $7{,}838{,}208a^5b^4$
4. $-8064x^5$
5. $\dfrac{1024}{59{,}049}$
6. $-489{,}888x^4$

9·5

1. $\frac{13}{204}$
2. $\frac{19}{55}$
3. $\frac{23}{42}$
4. No. p(student is a member of the orchestra) $\times$ p(student is a member of the honor society) $= \left(\frac{55}{100}\right)\left(\frac{42}{100}\right)$ $= \frac{231}{1000}$ does not equal p(student is a member of the orchestra and honor society) $= \frac{19}{100}$.
5. $\frac{20}{72} = \frac{5}{18}$
6. $\frac{32}{88} = \frac{4}{11}$
7. No. p(person is between 21 and 24) $\times$ p(person listens to music 2–3 hr per day) $= \left(\frac{78}{223}\right)\left(\frac{79}{223}\right)$ $= \frac{6162}{49{,}729}$ does not equal p(person is between 21 and 24 and listens to music 2–3 hr per day) $= \frac{34}{223}$.

9·6

1. $P(r=3) = \binom{6}{3}\left(\frac{1}{3}\right)^3\left(\frac{2}{3}\right)^3 = \frac{160}{729}$
2. $P(r \le 2) = \binom{5}{0}\left(\frac{1}{2}\right)^0\left(\frac{1}{2}\right)^5 + \binom{5}{1}\left(\frac{1}{2}\right)^1\left(\frac{1}{2}\right)^4 + \binom{5}{2}\left(\frac{1}{2}\right)^2\left(\frac{1}{2}\right)^3 = \frac{1}{2}$
3. $P(r \ge 2) = 1 - P(r \le 1) = 1 - \left(\binom{10}{0}\left(\frac{1}{6}\right)^0\left(\frac{5}{6}\right)^{10} + \binom{10}{1}\left(\frac{1}{6}\right)^1\left(\frac{5}{6}\right)^9\right) = \frac{10{,}389{,}767}{20{,}155{,}392}$
4. a. $P(p=100) = \binom{100}{100}(.99)^{100}(.01)^0 = 0.3660$

   b. $P(f=2) = \binom{100}{2}(.01)^2(.99)^{98}$
5. a. $P(r=5) = \binom{10}{5}(.78)^5(.22)^5 = 0.0375$

   b. $P(r \le 8) = 1 - P(r \ge 9) = 1 - \left(\binom{10}{9}(.78)^9(.22)^1 + \binom{10}{10}(.78)^{10}(.22)^0\right) = 0.6815$

   c. $P(r \ge 9) = P(r=9) + P(r=10) = \binom{10}{9}(.78)^9(.22)^1 + \binom{10}{10}(.78)^{10}(.22)^0 = 0.3185$

10 Introduction to statistics

10·1

1. Mean = 12.49 mm; median = 11.8 mm
2. Mean = 10.358 sec; median = 10.305 sec
3. Mean = 80.086; median = 80
4. Mean = 996.335 hr; median = 1000 hr

10·2

1. IQR: 15 mi; sample standard deviation: 10.4 mi
2. a. Mean: 71,956.3; median: 69,885

   b. IQR: 5967; sample standard deviation: 5758.4

3. a. Mean: 456.2 calories; median: 415 calories
   b. IQR: 160 calories; population standard deviation: 157.2 calories
4. The calories from McDonald's sandwiches have a lower measure of center than do the sandwiches offered by Burger King. There is less variation for the sandwiches offered by McDonald's. Although McDonald's has three sandwiches which constitute outliers in the spread of calorie ranges, these sandwiches have fewer calories than the highest calorie-rated sandwich from Burger King.

10·3

1. $P(61 < h < 63.5) = 0.341$
2. $P(56 < h < 66) = 0.954$
3. $P(h > 63.5) = .5 - P(61 < h < 63.5) = .5 - 0.341 = .159$
4. $P(h < 66) = .5 - P(61 < h < 66) = 0.5 + 0.476 = 0.977$
5. $P(h < 56 \text{ or } h > 66) = 1 - P(56 < h < 66) = 1 - 0.954 = 0.046$ (see question 2)
6. $P(h < 58.5 \text{ or } h > 66) = 1 - P(58.5 < h < 66) = 1 - 0.819 = 0.181$
7. $P(11.7 < \text{volume} < 12.2) = 0.5205$
8. $P(\text{volume} < 12.1) = 0.7161$
9. $P(\text{volume} > 12.4) = 0.0766$
10. $P(\text{volume} < 11.65 \text{ or volume} > 12.35) = 0.3368$
11. $P(t > 4.5) = 0.0401$

10·4

1. a.

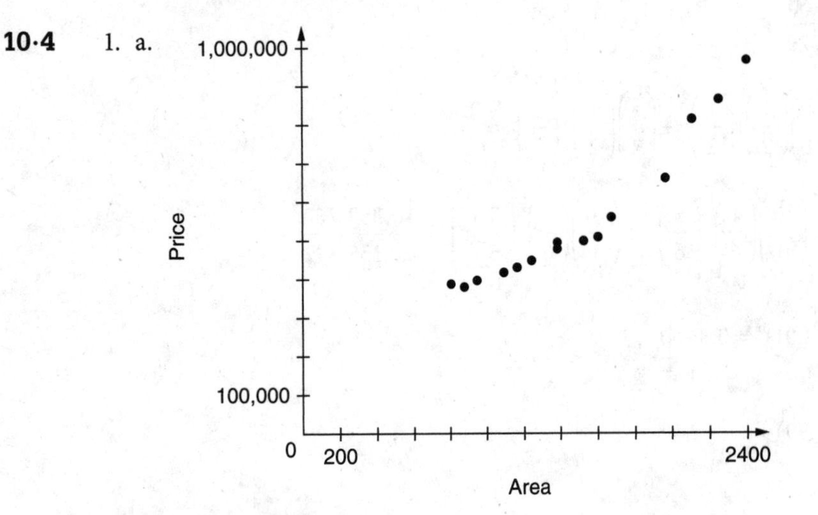

   b. Yes, $r = 0.9816$
   c. $\text{Price} = 124{,}364 \times (1.00085)^{\text{Area}}$
   d. The price of the house multiplies by a factor of 1.00085 for each additional square foot of area.
   e. Using the equation in part c, one would expect to pay about $624,859.

2. a.

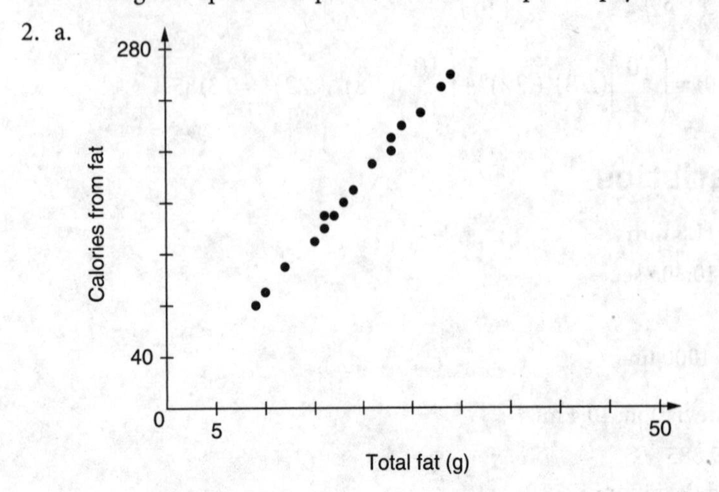

b. Yes, $r = 0.999097$

c. Calories = $9.021 \times \text{fat} - 1.322$

d. The number of calories in the sandwich increases by about 9.021 for each additional gram of fat in the sandwich.

e. 179.1 calories

3. a.

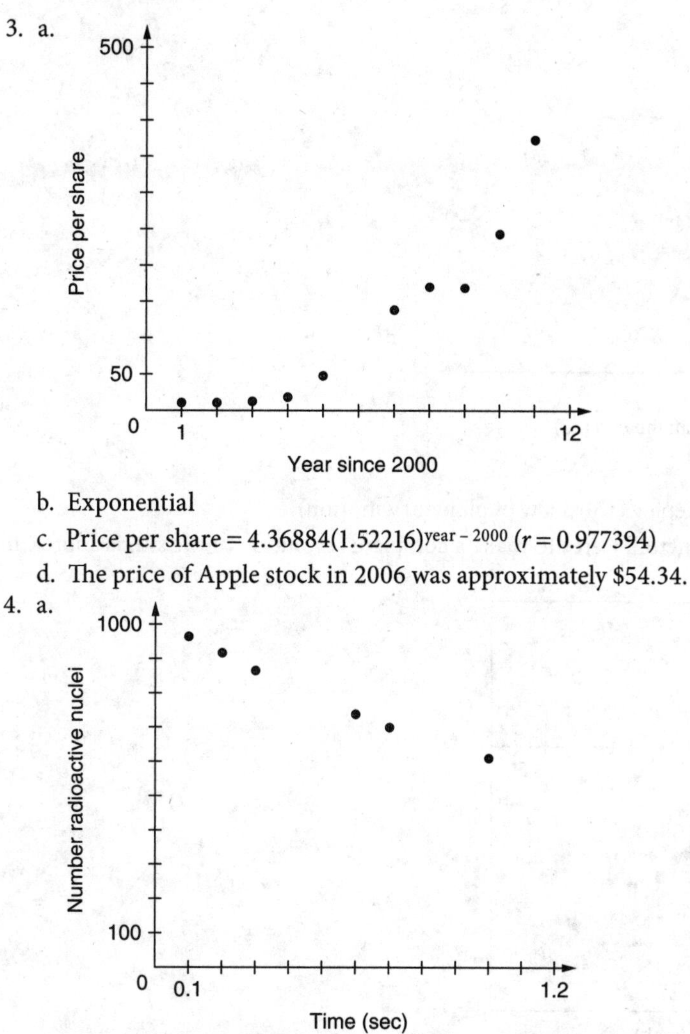

b. Exponential

c. Price per share = $4.36884(1.52216)^{\text{year} - 2000}$ ($r = 0.977394$)

d. The price of Apple stock in 2006 was approximately \$54.34.

4. a.

b. The correlation coefficient for the exponential model is -0.99929, while the correlation coefficient for the linear model is -0.995451. The data are best supported by an exponential model.

c. Number = $1010.01(0.595621)^{\text{time}}$

d. The number of radioactive nuclei after 0.5 sec is approximately 779.

5. a.

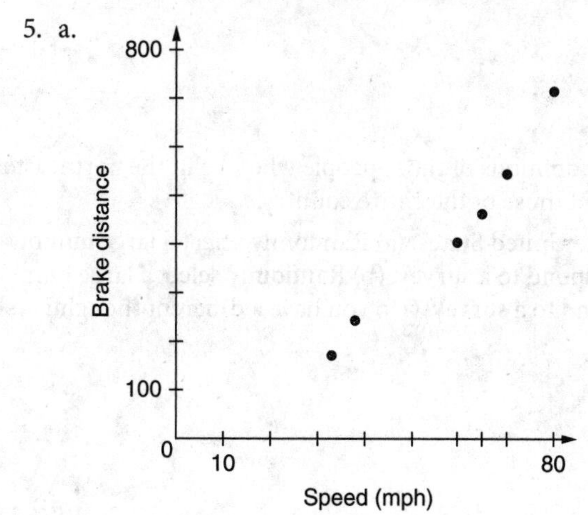

b. Power function

c. Distance = $0.1116 \times \text{speed}^{1.9997}$

d. The stopping distance on a wet road with tires having poor tread from a speed of 58 mph is about 375 ft.

6. a.

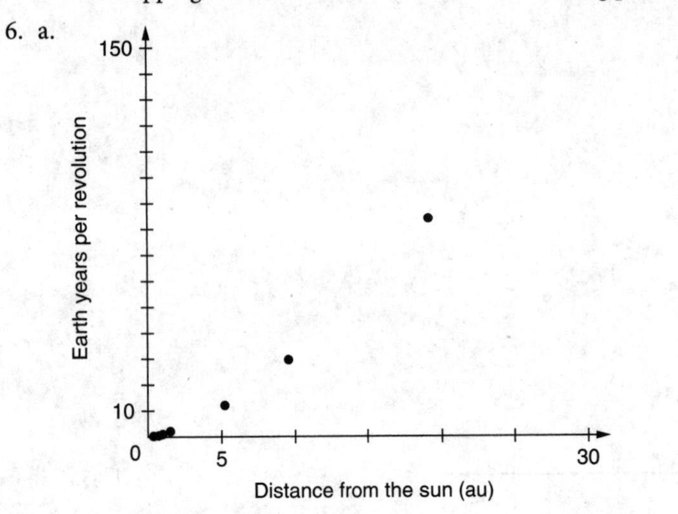

b. Power

c. yr = $0.999988 \times \text{dist}^{1.50036}$ (Kepler's third law of planetary motion)

d. The time needed for the planetoid Ceres to make a complete revolution of the sun is approximately 4.438 years.

7. a.

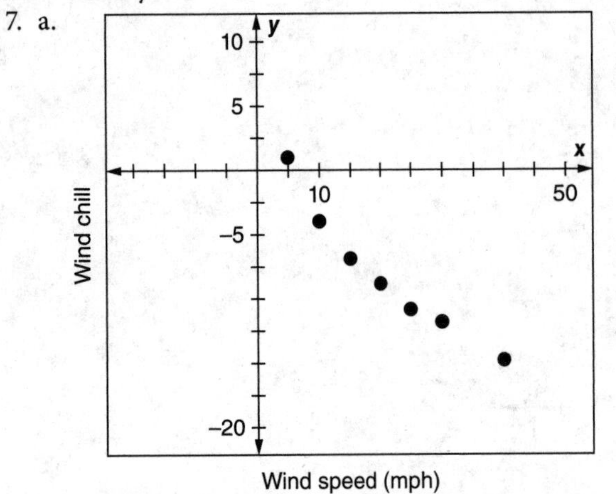

b. Logarithmic

c. Wind chill = 13.349 – 7.5506 ln(wind speed)

d. The wind chill when the air temperature is 10°F and the wind is blowing at 35 mph is approximately −13.5°F.

11 Inferential statistics

11·1

1. The survey is biased because of geography. The opinions of those people who live in the northeastern part of the United States do not necessarily represent those of the entire country.
2. Access a database for all registered voters in the United States. (a) Randomly select a large number of voters from this set of data and ask them to respond to a survey. (b) Randomly select a large number of voters within each state and ask them to respond to a survey. (Do you have a different thought? Ask your teacher to evaluate your response.)

11·2

1. 0.95
2. 1.28

3. 0.041
4. 0.018

For questions 5–7, the mean = 67.2 and the standard error of the mean $\frac{3.4}{\sqrt{50}} = 0.481$.

5. $P(66.5 < \bar{x} < 67.9) = 0.855$
6. $P(\bar{x} < 68.5) = 0.5 + P(67.2 < \bar{x} < 68.5) = 0.997$
7. $P(\bar{x} > 66) = 0.5 + P(66 < \bar{x} < 67.2) = 0.994$

For questions 8–10, the proportion = 0.98 and the standard error = $\sqrt{\frac{(0.98)(0.02)}{500}} = 0.00626$.

8. $P(0.975 < p < 0.99) = 0.733$
9. $P(p > 0.995) = 0.5 - P(0.98 < p < 0.995) = 0.008$
10. P(Less than 1% fail) = P(more than 99% meet manufacturer's specs) = $0.5 - P(0.98 < p < 0.99) = 0.055$

11·3

1. 1.75
2. −0.75
3. 1.6
4. A = 80.02, B = 94.38
5. A = 825.14, B = 982.86
6. invNorm(0.01, 1.002, 0.03) = 0.93 liters
7. invNorm(0.96, 16, 1.2) = 18.1 hands

11·4

1. $340.48 to $363.88
2. 11,259 to 12,141
3. 32.41% to 57.59%

11·5

1. H_0: The mean is greater than or equal to 1.01
   H_a: The mean is less than 1.01

   Critical Point: $invNorm\left(0.05, 1.01, \frac{0.02}{\sqrt{50}}\right) = 1.005$

   Decision: Fail to reject the null hypothesis
   Interpretation: There is no evidence to support the claim that the mean weight is different from the manufacturer's claim.
2. H_0: The proportion of dentists who recommend sugarless gum is 0.6
   H_a: The proportion of dentists who recommend sugarless gum is less than 0.6

   Critical Point: $invNorm\left(0.03, 0.6, \sqrt{\frac{(0.6)(0.4)}{400}}\right) = 0.5539$

   Decision: The proportion of dentists who recommend sugarless gum is 0.5625. Because this point is not inside the critical region, fail to reject the null hypothesis.
   Interpretation: There is no evidence to suggest that less than 60% of the dentists recommend sugarless gum.
3. H_0: The mean number of overtime hours each week is less than or equal to 40.4
   H_a: The mean number of overtime hours each week is greater than 40.4

   Critical Point: $invNorm\left(0.98, 40.4, \frac{3.8}{\sqrt{35}}\right) = 41.72$

   Decision: Reject the null hypothesis
   Interpretation: The data supports management's claim that the number of overtime hours worked each week is increasing.

4. H_0: The proportion of rock-and-roll love songs played on the radio is greater than or equal to 0.5
H_a: The proportion of rock-and-roll love songs played on the radio is less than 0.5

Critical Point: $invNorm\left(0.01, 0.5, \sqrt{\frac{(0.5)(0.5)}{125}}\right) = 0.3960$

Decision: Because 48 of the 125 songs, or 38.4%, were "love" songs, reject the null hypothesis.
Interpretation: The evidence supports Jack's claim that less than 50% of the rock-and-roll songs played on the radio are "love" songs.

11·6

1. Designate the numbers 0–4 to represent the different dinosaurs. Use a random number generator (such as randint(0,4,n) where n represents the number of boxes of cereal purchased) to determine the number of boxes of cereal that need to be purchased before the consumer gets one of each of the values 0 through 4. Repeat to get a large number of trials.
2. Designate the numbers 0–54 to indicate that the Western team wins a game and the numbers 55–99 to indicate that the Eastern team wins. Each trial can be randint(0,99,7). Read from left to right to determine which group reaches four wins first. Repeat to get a large number of trials.

12 Trigonometry: Right triangles and radian measure

12·1

1. 74.2
2. 72.5
3. 171.6
4. 17
5. 45
6. 41
7. 202

12·2

1. 2
2. $\frac{2}{\sqrt{3}} = \frac{2\sqrt{3}}{3}$
3. $\sqrt{3}$
4. $\sqrt{2}$
5. $\sqrt{2}$
6. 1
7. $\frac{2}{\sqrt{3}} = \frac{2\sqrt{3}}{3}$
8. 2
9. $\frac{1}{\sqrt{3}} = \frac{\sqrt{3}}{3}$
10. $\frac{25}{7}$
11. $\frac{24}{25}$
12. $\frac{25}{24}$
13. $\frac{24}{7}$
14. $\frac{7}{24}$
15. $\frac{8}{9}$
16. $\frac{\sqrt{17}}{9}$
17. $\frac{9}{\sqrt{17}} = \frac{9\sqrt{17}}{17}$
18. $\frac{8}{\sqrt{17}} = \frac{8\sqrt{17}}{17}$
19. $\frac{\sqrt{17}}{8}$

12·3

1. 479, −241
2. 77, −283
3. 263, −457
4. 85
5. 10
6. 70
7. 23
8. tan(37°)
9. −sin(41°)
10. −cos(51°)
11. $\left(\frac{-1}{2},\frac{\sqrt{3}}{2}\right)$
12. $\left(\frac{-\sqrt{2}}{2},\frac{-\sqrt{2}}{2}\right)$
13. $\left(\frac{\sqrt{3}}{2},\frac{-1}{2}\right)$
14. $\frac{-12}{13}$
15. $\frac{-40}{9}$

12·4

1. $\frac{2\pi}{5}$
2. $\frac{7\pi}{9}$
3. $\frac{7\pi}{4}$
4. 40°
5. 50°
6. 75°
7. 2.5
8. $\frac{20}{3\pi}$

12·5

1. $\frac{-56}{65}$
2. $\frac{-33}{65}$
3. $\frac{56}{33}$
4. $\frac{-49}{71}$
5. $\frac{60+11\sqrt{3}}{122}$
6. $\frac{-60-11\sqrt{3}}{122}$
7. $\frac{-3+\sqrt{21}}{8}$
8. $\frac{3\sqrt{3}-\sqrt{7}}{8}$
9. $8+3\sqrt{7}$

12·6

1. 219 in^2
2. 3019 ft^2
3. 75° or 105°
4. 141°

12·7

1. 28.6°
2. 36.0°
3. $XZ = 40.2$ cm, $YZ = 50.6$ cm
4. $QS = 8.1$ cm, $RS = 7.2$ cm
5. $ES = 47.1$ cm, $EY = 55.3$ cm
6. $QE = 42.7$ cm, $ED = 56.5$ cm

12·8

1. 1
2. 1
3. 1 (82 > 44)
4. 1 (length of the side opposite the obtuse angle is greater than the length of the adjacent side)
5. 0 (length of the side opposite the obtuse angle is less than the length of the adjacent side)
6. 2 (length of the opposite side is greater than the minimum length 72 sin(50) ≈ 55.16, but less than 72)
7. 2 (length of the opposite side is greater than the minimum length 125 sin(57) ≈ 104.83, but less than 125)

12·9

1. 39 cm
2. 115 ft
3. 44.5°
4. 112.8°
5. 148.5 N
6. 87°
7. 36.0°

13 Graphs of trigonometric functions

13·1

1. 8
2. 4
3. $f(19) = f(3) = 4$
4. $f(-203) = f(5) = -4$
5. 8. The graph is stretched from the x-axis by a factor of 2 and then translated down 1 unit.

   The range of $f(x)$ is $-4 \le y \le 4$, while the range of $g(x)$ will be $-9 \le y \le 7$. The amplitude is $\frac{7-(-9)}{2} = 8$.

13·2

1. The graph of the sine function is translated down 1 unit.

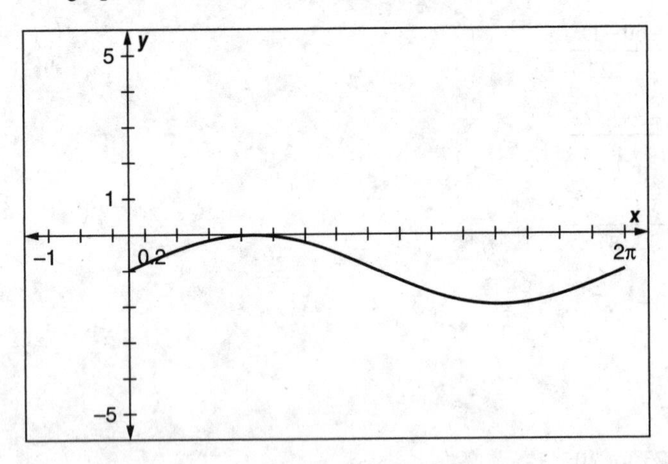

2. The graph of the cosine function is translated up 1 unit.

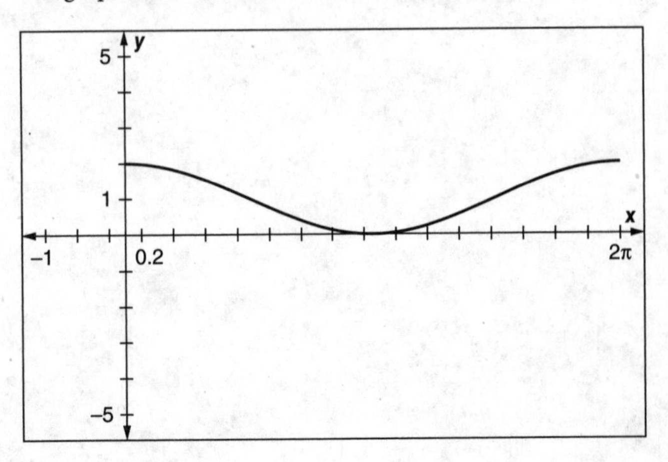

3. The graph of the sine function is translated right $\frac{\pi}{2}$ unit.

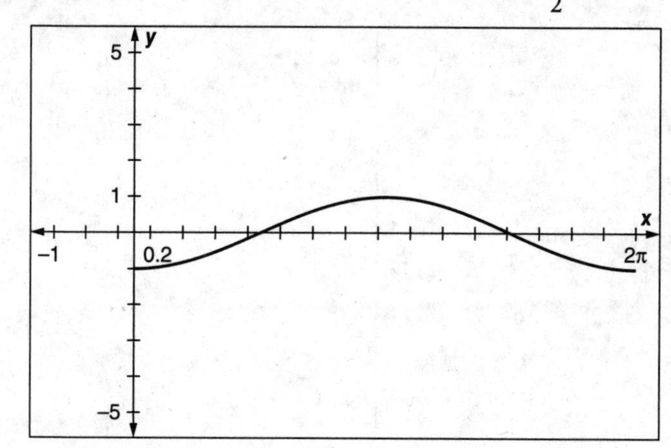

4. The graph of the cosine function is translated left $\frac{\pi}{3}$ unit.

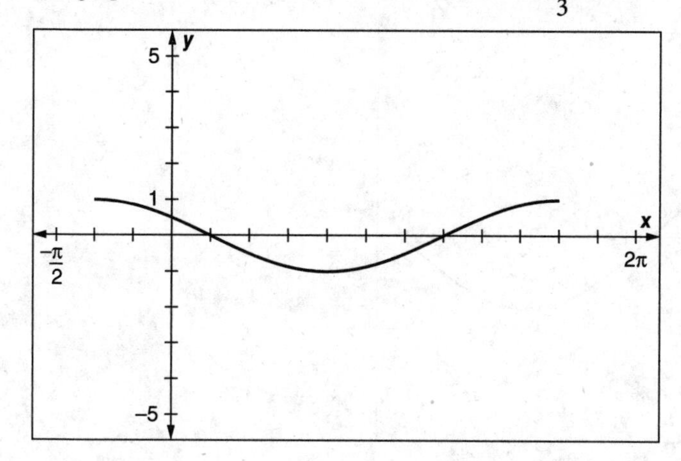

5. The graph of the tangent function is translated right $\frac{\pi}{4}$ unit.

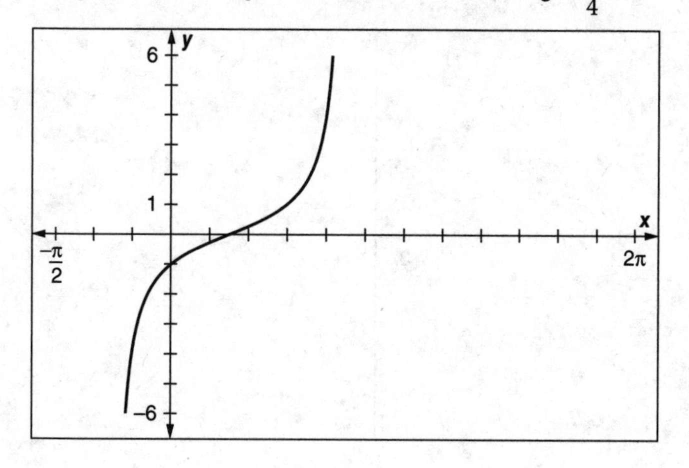

6. The graph of the tangent function is translated down 1 unit.

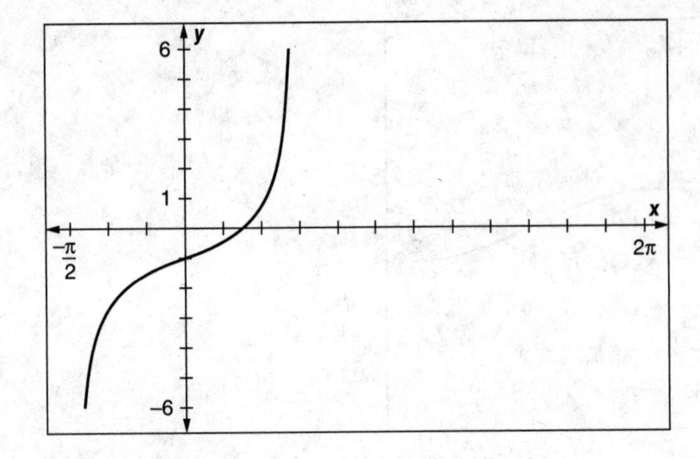

7. The graph of the sine function is stretched from the x-axis by a factor of 2.

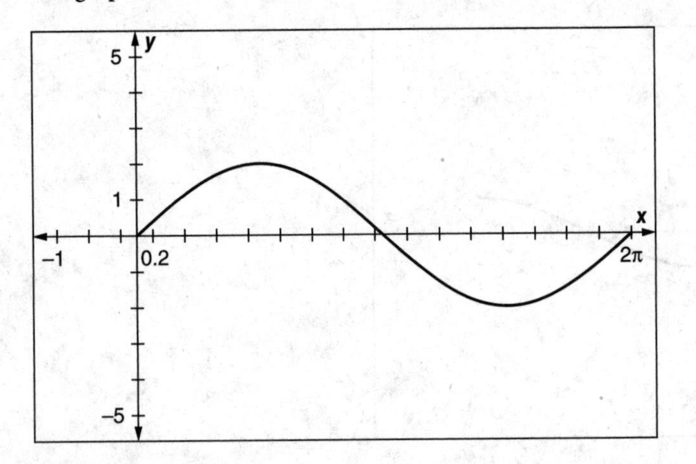

8. The graph of the sine function is stretched from the y-axis by a factor of $\frac{1}{2}$.

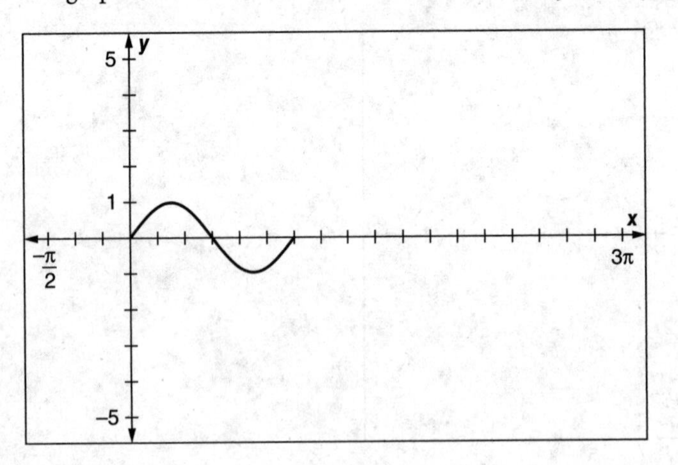

9. The graph of the cosine function is stretched from the y-axis by a factor of $\frac{1}{\pi}$.

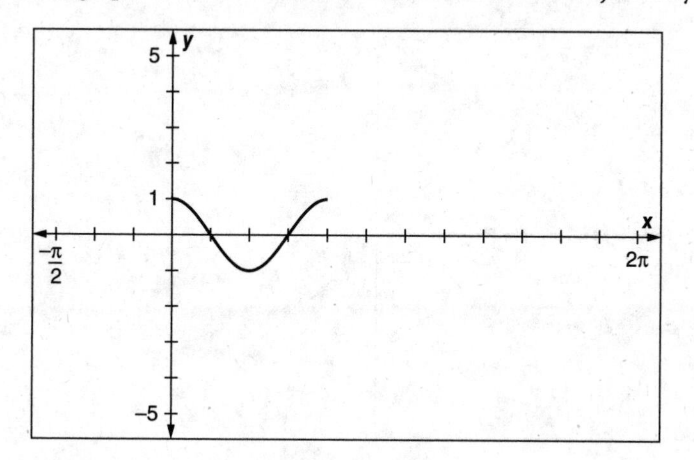

10. The graph of the cosine function is stretched from the y-axis by a factor of $\frac{1}{2}$, stretched from the x-axis by a factor of 2, and translated down 1 unit.

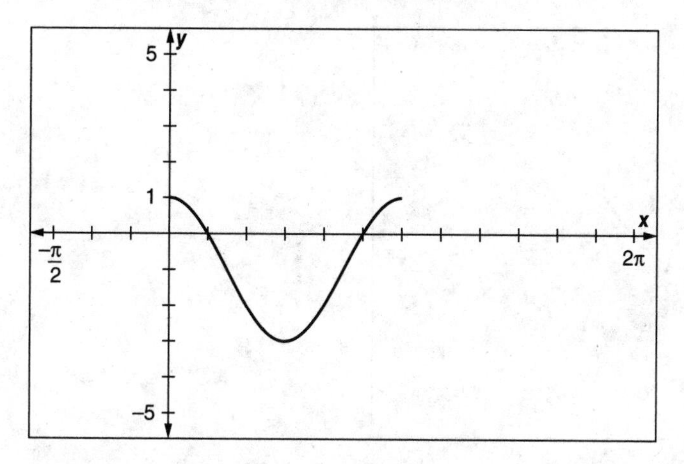

11. The graph of the sine function is translated right $\frac{\pi}{2}$, reflected over the x-axis, and stretched from the x-axis by a factor of $\frac{1}{2}$.

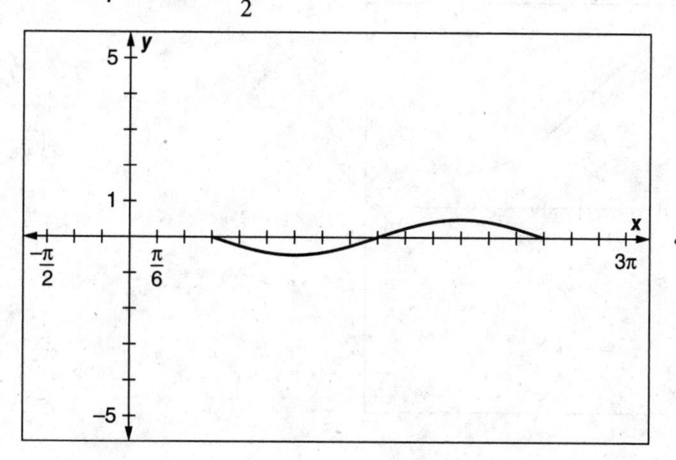

12. The graph of the tangent function is stretched from the y-axis by a factor of $\frac{1}{2}$.

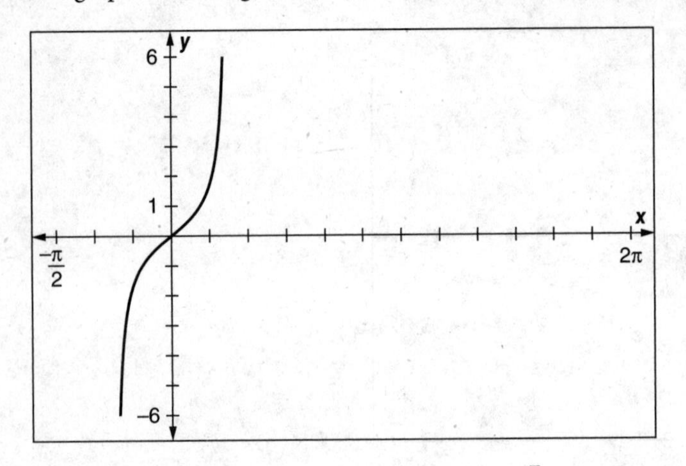

13. The graph of the sine function is translated left $\frac{\pi}{4}$, stretched from the x-axis by a factor of 2, and translated up 1.

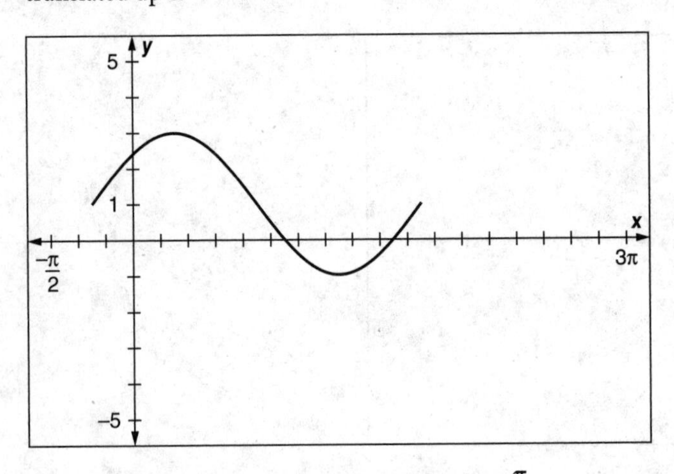

14. The graph of the cosine function is shifted left $\frac{\pi}{3}$, stretched from the x-axis by a factor of 2, reflected over the x-axis, and translated up 3 units.

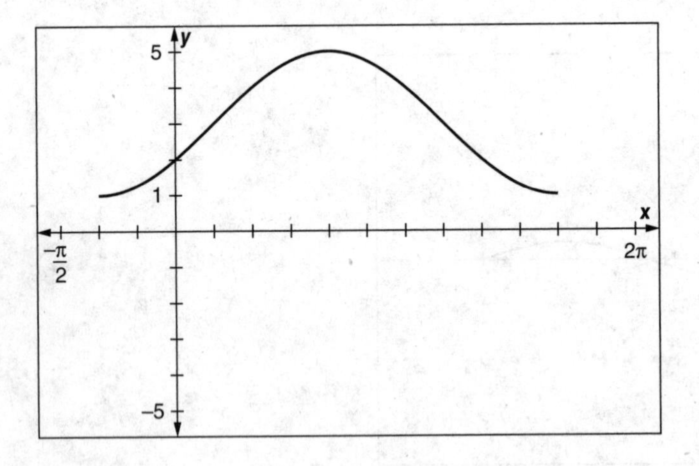

15. Amplitude: 2; period: 2π; phase shift: left $\frac{\pi}{4}$; vertical translation: up 1

16. Amplitude: 2; period: 2π; phase shift: right $\frac{\pi}{6}$; vertical translation: up 3. (An acceptable answer would be a phase shift left by $5\pi/6$, or right by $7\pi/6$. Since multiplying cos by -1 induces a phase shift right by π, on top of a phase shift right by $\pi/6$, you get a phase shift right by $7\pi/6$, or a phase shift left by $5\pi/6$.)

17. Because the period of the sine function is 2π, $\sin(z) = \sin(z + 2\pi) = h$.

18. Because the period of the cosine function is 2π, $\cos(z) = \cos(z + 2\pi)$. However, $\cos(z + 3\pi)$ will be another half revolution about the cycle, so $\cos(z + 3\pi) = -k$.

19. $\frac{43.95 - 6.85}{2} = 18.55$ ft

20. 25.4 ft

21. $h = 18.55\cos\left(\frac{4\pi}{25}t\right) + 25.4$

13·3

1. $\frac{-\sqrt{55}}{11}$

2. $\frac{-5}{\sqrt{74}} = \frac{-5\sqrt{74}}{74}$

3. $\frac{9 - x^2}{x^2 + 9}$

4. $\frac{21}{29}$

5. $\frac{5\sqrt{3} + 2}{12}$

13·4

1. 228.6°, 311.4°
2. 60.9°, 240.9°
3. 0°, 60°, 300°
4. 19.5°, 160.5°
5. 16.6°, 124.8°, 196.6°, 304.8°
6. 0°, 60°, 180°, 300°
7. 38.7°, 218.7°
8. 23.0°, 157.0°, 219.8°, 320.2°
9. 0°, 120°, 240°
10. 0°, 45°, 135°, 180°, 225°, 315°
11. 2.6 and 9.9 hr after high tide